T0255803

AI FOR CARS

AI FOR EVERYTHING

Artificial intelligence (AI) is all around us. From driverless cars to game-winning computers to fraud protection, AI is already involved in many aspects of life, and its impact will only continue to grow in the future. Many of the world's most valuable companies are investing heavily in AI research and development, and not a day goes by without news of cutting-edge breakthroughs in AI and robotics.

The *AI for Everything* series will explore the role of AI in contemporary life, from cars and aircraft to medicine, education, fashion and beyond. Concise and accessible, each book is written by an expert in the field and will bring the study and reality of AI to a broad readership including interested professionals, students, researchers and lay readers.

AI for Immunology
Louis J. Catania

AI for Cars
Josep Aulinas & Hanky Sjafrie

AI for Digital Warfare
Niklas Hageback & Daniel Hedblom

AI for Fashion
Anthony Brew, Ralf Herbrich,
Christopher Gandrud, Roland Vollgraf,
Reza Shirvany & Ana Peleteiro Ramallo

AI for Death and Dying
Maggi Savin-Baden

AI for Radiology
Oge Marques

AI for Games
Ian Millington

AI for School Teachers
Rose Luckin & Karine George

AI for Learners
Carmel Kent, Benedict du Boulay &
Rose Luckin

AI for Social Justice
Alan Dix and Clara Crivellaro

For more information about this series please visit:
https://www.routledge.com/AI-for-Everything/book-series/AIFE

AI FOR CARS

Josep Aulinas and Hanky Sjafrie

CRC Press
Taylor & Francis Group
Boca Raton London New York

CRC Press is an imprint of the
Taylor & Francis Group, an **informa** business
A CHAPMAN & HALL BOOK

First edition published 2022
by CRC Press
6000 Broken Sound Parkway NW, Suite 300, Boca Raton, FL 33487-2742

and by CRC Press
2 Park Square, Milton Park, Abingdon, Oxon, OX14 4RN

CRC Press is an imprint of Taylor & Francis Group, LLC

Library of Congress Cataloging-in-Publication Data

ISBN: 978-0-367-56830-6 (hbk)
ISBN: 978-0-367-56519-0 (pbk)
ISBN: 978-1-003-09951-2 (ebk)

DOI: 10.1201/9781003099512

Typeset in Joanna
by MPS Limited, Dehradun

CONTENTS

FOREWORD

AI is transforming virtually every industry from healthcare to finance to entertainment. However, perhaps no industry is poised for the biggest revolution than the transportation industry. AI and its resulting autonomous vehicles will no doubt make our roads much safer and more efficient – and ultimately alter how roads, cities and suburbs are designed.

Creating automated and autonomous vehicles is a massive technical challenge, likely one of the world's most complex. It is not just about gathering vast sums of data from cameras, radars and lidars, and performing trillions of calculations in a fraction of a second. Lives are on the line, and this entire process must be performed to deliver the highest level of safety possible. The transformation also encompasses a shift from fixed-function systems inside the vehicle, to a true software-defined platform that is perpetually updateable.

The challenge extends from end to end – from the datacenter where the AI is first trained, through a comprehensive suite of tests to validate the system, to the diverse and redundant systems inside the vehicle. AI can be utilized at all stages of the real-time computing pipeline, from perception (both outside and inside the car), to mapping and localization, to planning and control. And the process never ends. Like humans, who never stop learning, the AI car will get smarter and smarter over time as the software is trained for new tasks, enhanced, tested and validated, then updated over the air.

AI for Cars breaks down this challenge into a myriad of individual building blocks, providing insights throughout. From pedestrian detection to driver monitoring to recommendation engines, the book covers the background, research and progress thousands of talented engineers and researchers have achieved thus far, and their plans to deploy this life-saving technology all over the world.

One common theme appears throughout the book: autonomous vehicles are a computational challenge of astronomical proportions. It is no longer about the engine's horsepower, but rather the vehicle's computing horsepower. Greater computational performance in the datacenter accelerates training of the wide array of deep neural networks that ultimately will run inside the vehicle, and also runs the simulation technology required to test and validate the dangerous and challenging scenarios without putting anyone in harm's way. Inside the vehicle, greater computational performance enables the use of higher resolution sensors, as well as more diverse and redundant algorithms. Greater computational performance equals greater safety.

Enjoy the ride.

Danny Shapiro,
NVIDIA

PREFACE

The emerging field of Artificial Intelligence has had its share of advocates and detractors. For some people, the advent of AI is the harbinger of a coming techno-paradise; others see it as a Trojan horse: a dreaded sign of the ultimate takeover of humanity by nefarious androids.

As utopian or foreboding as these visions of the future may be, we may already have been using AI technology several times daily without realizing it – and without many complaints. AI's basic assumption that machines can be made to think and make decisions like human beings can be scary [1]. But another assumption that goes with that one is that these machines can be very helpful when they are thus enabled. Asking our smart speaker for tomorrow's weather forecast utilizes AI, as does choosing an automatic blur or virtual background for video conferencing. AI is also at work when product recommendations appear to us online as a result of our purchase history.

Although the era of the automobile began well before the computer age, the latter's progress has been faster. Not many decades after a few monstrous, room-sized computers lumbered off the starting blocks in the first computer era, teenagers now slip powerful computing devices into their pockets without a second thought and the fourth computer era (including self-learning by our increasingly capable electronic counterparts) is just around the corner. But far from threatening to replace the automobile, the computer now animates it; modern cars are in a state of ongoing transformation into increasingly software-driven products. Also, most automotive innovations involve software to both process information and communicate the results to the driver – and that trend is increasing [2]. In this environment, AI software holds the potential to bring automotive innovation to the next level.

Beyond in-car applications, AI technology has also been deployed in numerous other automotive domains. These include supply-chain optimization (which we may have expected), fully automated robots in production lines (the assembly line worker's presumptive dream and nightmare) and even AI-powered augmented reality in vehicle design.

Whether surprising or not, these examples illustrate the fact that AI is starting to drive development in many areas. To help us all – including tech fans – make sense of such a varied backdrop, this book aims to give some insight into the AI applications being employed in the automotive field by showcasing selected real-world use cases. Concerning how we drive our cars, we will look at how AI can increase road safety in the contexts of advanced driver assistance systems and in-vehicle infotainment systems; in addition, we will look ahead to some features of autonomous driving. Not leaving out the background, we will also consider recent developments in research and development (virtual testing and the generation of synthetic scenarios) and in car services (including predictive diagnostics and maintenance and even driver behavior analysis). Lastly, we will also have a look at AI in car safety and car security.

AI technology has already become not only ubiquitous but highly useful, making our modern lives function better by providing new and faster ways to do things we want to do – be it at the workplace or in our hobbies or quotidian chores. Whether we see these developments as positive or negative, there's no denying that we live in a time of high-paced technological advances in which AI plays a growing role.

With this book, we'd like to invite you on a short guided tour through many different AI landscapes including robotics, image and speech processing, recommender systems and on to deep learning, all within the automobile world. In fact, cars inhabit one of just a few domains where you can find so many AI innovations packed into a single product. So let's buckle up and get started.

1

AI FOR ADVANCED DRIVER ASSISTANCE SYSTEMS

Advanced Driver Assistance Systems (ADAS) can be defined as a collection of electronic systems that aid drivers by improving safety, comfort and efficiency while driving and parking. Even if you've never heard the term ADAS before, it's quite possible that you – as a car user or owner – have been using some ADAS functions, for instance, an Anti-lock Braking System (ABS) or Electronic Stability Control (ESC), without noticing. ABS and ESC are just two popular examples of many ADAS applications that find their ways from niche innovations to standard car safety features.

In contrast to AD, which is generally placed at one of the levels 3–5 (conditional, high or full automation) on the scale created by the Society for Automotive Engineers (SAE) International, ADAS is commonly located at SAE level 2 (partial automation). This means that the driver is assisted by ADAS but remains the primary actor controlling the vehicle.

ADAS functions vary from purely informative, such as speed limit information, to safety-critical ones, such airbag deployment.

Similar to safety-related systems in other domains such as aerospace or medical devices, the whole development process of ADAS needs to comply with industry-wide standards as well as other governing regulations. One such standard is the ISO 26262 "Road Vehicles – Functional Safety" standard [3]. ISO's Automotive Safety Integrity Levels (ASILs) for each ADAS function are determined according to the levels of severity, exposure and controllability of hazardous events that might arise from the system's malfunction. The ISO 26262 defines four ASIL levels, ranging from ASIL A (the least stringent level) to ASIL D (the most stringent level). As such, ASIL D functions require more comprehensive safety requirements and measures than ASIL A, B and C. Functional safety and ISO 26262 will be covered in more detail in the last chapter of this book.

Developing safe ADAS functions that work reliably everywhere is undoubtedly challenging. In the age of digital connectivity, we're living in, however, they are not the only variables in the equation. For ADAS to promote safety as intended, a car's systems must be securely protected from unwanted influences and even cyber-attacks. The famous "Jeep attack" case showed how serious the consequences of unaddressed security holes (which the automaker Chrysler later corrected) can be [4]. We will discuss car security and some of the AI-driven cybersecurity measures toward the end of this book.

Other challenges in ADAS include the necessity of reliable functionality wherever motor vehicles may be used. Traffic laws and signs are not globally standardized; in addition, there can be temporary and long-term changes in speed limits and other posted restrictions due to construction and spontaneously arising conditions such as accidents and weather. ADAS must also be able to work dependably in widely differing and sometimes very harsh physical environments, ranging from desert heat to arctic winters.

Yet in the face of all these challenges, AI's contribution to the progress in ADAS is already making driving both easier and safer. To look a bit more closely at these practical benefits, in this chapter we will consider the advances made with AI's help in three ADAS applications, namely automatic parking, traffic sign recognition and Driver Monitoring Systems (DMS).

AUTOMATIC PARKING

For a lot of people, parking is probably the least enjoyable aspect of driving a car. Finding an empty parking space in big cities can be challenging, and trying to park in a tight slot on a busy street can be downright stressful. But maybe especially in less stressful cases, parking assistant systems can be very helpful.

Over the past two decades, parking assistant systems have evolved from purely informational to fully automatic systems. Whereas the early versions of such systems could only visualize the rear environment of the vehicle with the help of a rear-view or backup camera, modern automatic parking systems park and unpark the vehicle autonomously without requiring any human inside the vehicle.

The transition from parking assistant systems (which were non-automatic) to semi-automatic parking systems for series-production cars was pioneered by Toyota through the introduction of Intelligent Parking Assist in its Prius hybrid model in Japan in 2003. The initial version of the system supported only the reverse parallel parking scenario. The required parking maneuver was automatically calculated by the system, so that the driver could take their hands off the steering wheel and control the maneuver using the brake pedal. The introduction of remote-control parking in the BMW 7 Series and the Mercedes E-Class during the years 2015/2016 marked the transition from semi-automatic to fully automatic systems for series-production cars. This automatic parking system enabled a vehicle to park and unpark itself autonomously within a parking lot or a garage without requiring any human assistance from inside the vehicle. The user just needed to be in proximity to the car, to control the automatic parking maneuver by pressing a button in the car key or in a smartphone app. In 2019, an automated valet parking system in Stuttgart, Germany became the first system of its kind to receive official approval for daily use from the local transport authority [5]. Using the automated valet parking system, the driver can just leave the vehicle at the drop-off area, after which it will park itself inside the intelligent parking building without any human involvement. Likewise, the vehicle can unpark and navigate itself back to the drop-off area when the owner summons it.

Modern automatic parking systems typically support common parking scenarios. These especially include parallel parking (where the parking space is parallel to the road), angle parking (where the space is arranged at an acute angle – smaller than 90° – to the road) and perpendicular parking (where the space is situated perpendicular – at a 90° angle – to the road).

In terms of automation level, parking systems can be generally divided into two categories: semi-automatic and fully automatic. Semi-automatic parking systems require the driver to press the gas pedal to start or proceed with the parking maneuver, shift gears (into forward or reverse) and if necessary, press the brake pedal to interrupt the maneuver. The system controls the steering wheel automatically, based on the calculated trajectory for the parking space.

By contrast, fully automatic parking systems do not require manual control of the transmission nor of the gas and brake pedals, as the vehicle is capable of performing the parking maneuver autonomously from start to finish. For this reason, it is not even necessary for the driver to be inside the car while the vehicle is parking in a parking garage. However, for safety reasons, it is usually still required for the driver or an operator to keep pressing the "hold-to-run button" on the car key or on the smartphone app during the parking maneuver. If the button is no longer pressed, the parking maneuver stops immediately to prevent accidents from arising due to unattended operation. The automatic parking system is designed to be automatically interruptible whenever there could be a collision with humans or inanimate obstacles.

The more advanced form of fully automatic parking is automated valet parking. The concept is similar to conventional valet parking service, in which the driver and all passengers alight at a drop-off area and a person parks the vehicle for them and returns it when they need it back. The only difference in the automated valet parking scenario is that there is no valet. The vehicle parks autonomously, driving itself from the drop-off area into the parking space and later returning autonomously at the customer's request. Automated valet parking typically requires additional sensors or other equipment installed in the parking infrastructure to help vehicle navigation as well as to provide other important information for the vehicle, such as the location of a free parking space together with a high-definition map of the parking area. The additional "help" from the parking infrastructure is particularly important for automated valet parking in an indoor parking garage, since Global Navigation Satellite Systems (GNSS) such as Global Positioning System (GPS) are often not available to help with localization.

Automatic parking systems are made possible through well-orchestrated actions involving several ADAS functions such as parking space detection, parking spot marking lane detection, object detection, localization, path planning and path tracking. Depending on the vehicle architecture and the complexity of the parking scenarios, these systems require various amounts of data from multiple sensors and good coordination of several ADAS Electronic Control Units (ECUs).

For a vehicle to detect a suitable parking space according to present technology, the driver usually first activates the automatic parking system and slowly passes by a potential space. While passing, the onboard sensors scan the parking lot, measure the free space and determine whether the space is suitable for automatic parking. This step might not be necessary in the case of the automated valet parking scenario, as the parking infrastructure should be able to provide information about available parking spots directly to the vehicle. Although the ultrasonic sensor is the most common sensor type used for parking space detection, several studies have also showed promising results for parking space detection using a camera [6] or a combination of ultrasonic sensor and camera [7]. The lower cost of cameras and the richer information they provide make them an attractive sensor for parking space detection; however, their usage can be limited in poor lighting environments.

To ensure that vehicles are parked within the defined parking space boundaries, lane-marking detection is necessary. We should note here that this form of detection is also important for automatic parking scenarios when there is no adjacent vehicle. This is due to the fact that ultrasonic sensors can only sense when an ultrasonic wave is reflected by an adjacent object and returns to the sensor. If there is no adjacent object (like another car) and no other physical structure (such as walls) to help define the limits of the parking space, the vehicle might park anywhere – irrespective of the markings on the floor. For this reason, cameras are still the predominant sensors for lane-marking detection. Lidars have also been investigated for lane-marking detection, although at present their relatively high price tag still hinders the broad adoption of this technology in cars. Lidar-based detection works by analyzing the intensity of the point clouds reflected by the lane markings on the surface of the floor [8]. Laser beams reflected by markings have a different intensity than those from unmarked surfaces. Because lidar sensors work in all lighting conditions, they could become a viable option to compensate camera limitations in dark parking environments.

Throughout the parking maneuver, the vehicle's surrounding environment is permanently monitored with object detection to avoid collisions with any obstacles, whether static (such as walls, objects on the ground etc.) or moving (humans, animals etc.). Whereas path planning algorithms such as Rapidly-Exploring Random Trees (RRT) [29] and Hybrid A-Star (Hybrid A*) [9] calculate the necessary trajectory of a vehicle to reach the desired parking location, path tracking ensures that the vehicle follows the planned trajectory properly, using for instance the pure pursuit [10] or the Stanley method [11].

For automated valet parking use cases, the vehicle must also localize itself along the travel path from the drop-off point to the destination parking space or vice versa. Depending on the parking space environment (indoor/outdoor), the vehicle sensor configuration and the availability of additional sensors/equipment installed in the parking infrastructure, there are several approaches to perform this localization: combining the Inertial Navigation System (INS) and GNSS, using external references (beacons, magnets, visual markers etc.) or using the on-board lidar or camera sensor only [12]. In the latter case, the parking environment is previously mapped using a Simultaneous Localization and Mapping (SLAM) algorithm and the localization is then applied on the generated high-resolution map of the parking environment. SLAM algorithms can generally be divided into filtering and optimization approaches [13]. The filtering approach, for example, the Extended Kalman Filter SLAM (EKF-SLAM) [14] or the FastSLAM algorithm [15], summarizes information from past observations and iteratively (i.e. through repetition) improves its internal belief (its appraisal of what it has perceived through the car's sensors) by incorporating new observations as they are gathered. By contrast, the optimization approach, such as the graph-based SLAM algorithm [16], keeps track of all poses (the car's positions and the directions in which it has been pointed) and measurements from the beginning until the current observation and finds the most likely overall trajectory or the one most consistent with the whole observation set.

Automatic parking systems not only increase drivers' comfort level; they also have the potential to reduce traffic accidents and to result in more effective parking lot use in the case of automated valet parking. The latter is undoubtedly an attractive way to address parking space issues in cities and other areas experiencing a parking space shortage. However, the current technology for automated valet parking systems still requires some high-tech investment in the parking infrastructures and possibly also in the vehicles. Hence, the wide adoption of such systems depends not only on the maturity level of the technology but also on the economic aspects of its application.

TRAFFIC SIGN RECOGNITION

Have you ever missed or forgotten the last speed limit sign because you were so engaged in a discussion while driving or maybe were simply not attentive enough? Or have you ever driven on rural roads abroad while wondering what the actual speed limit is, since there was no speed limit sign to be seen?

Thanks to affordable personal navigation devices and the navigation apps on our smartphones nowadays, we may not be completely lost in these situations as our devices or apps could show us the valid speed limit throughout our journey. However, the reliability of this information is heavily dependent on the quality of the map material referred to and the availability of satellite navigation systems such as the GPS.

If the map material is outdated or incomplete, the speed limit shown might be incorrect – or entirely absent in the case of new roads. Speed limits might also not be shown correctly when one is driving through a long tunnel with more than one-speed limit posted inside it, since the navigation assistants would lose track of the vehicle's position due to the unavailability of the GPS signals.

Another drawback of map-based speed limit information is the lack of consideration of what are known as variable speed limits. Typically shown on electronic signs or portable signposts and only valid temporarily, these flexible restrictions take priority over the permanent ones on specific road segments.

Variable speed limits are used by local transportation or law enforcement authorities to regulate traffic flow, for instance, to avoid or reduce the impact of road congestion or to improve safety for road workers. Map-based speed limit information is not able to take these variations into account and will therefore show an incorrect speed limit.

To overcome the above limitations, modern cars typically use a combination of camera-based traffic sign recognition and map input to provide highly accurate speed limit information. In the case of variable speed limit situations, the camera-detected speed limit takes priority over the information embedded in the map material. On the other hand, the information from the map material will take priority if no speed limit signs are detected for a long time, or when camera recognition is temporarily unavailable or unreliable, e.g. due to glare or heavy rain.

Due to advances in ADAS, speed limit signs are far from being the only traffic signs that cars are able to recognize nowadays. AI has for example empowered vehicles to recognize no-entry/wrong-way, no-passing, yield and stop signs and even the color of traffic lights, from a distance of up to 150 meters [17]. The information is typically shown using pictograms displayed in the cluster-instrument, head-unit or head-up display. Traffic sign recognition is primarily done using cameras. Some studies have also proposed a sensor fusion of camera and lidar (light detection and ranging) to improve overall system accuracy [18], [19].

Accurate traffic sign recognition can potentially save lives, helping warn drivers or even keeping them from dangerous violations of traffic regulations caused by inattentiveness. To provide a strong incentive for car manufacturers to integrate this and other useful ADAS features such as Autonomous Emergency Braking (automatic braking upon detection of a likely collision) and Lane Keeping Assist (automatic directional correction in case of vehicle drift beyond lane boundary) to improve the safety of their products, the European New Car Assessment Programme (Euro NCAP) has included speed assistance systems as part of their safety rating since 2009 [20]. An international non-profit organization backed by several European governments, consumer groups and motoring clubs, Euro NCAP aims to provide independent safety assessment of new cars sold in Europe.

The Euro NCAP safety assessment is conducted in four areas: adult occupant, child occupant, vulnerable road user protection and safety assist. An overall safety rating ranging from zero to five stars is awarded if the car's individual scores exceed the minimum score for each area specified for that safety rating. For example, to achieve a five-star overall safety rating for the assessment year 2020–2022, the car must achieve the minimum score of 80% for the adult occupant area, 80% for the child occupant area, 60% for the vulnerable road user protection area, and 70% for the safety assist area [21]. For a five-star safety rating in the assessment year 2023–2024, the same minimum scores apply for the adult occupant, child occupant and safety assist areas, while the minimum score for the vulnerable road user protection area is increased to 70% [21].

Although traffic sign recognition was offered as an ADAS feature in cars starting in the late 2000s, research into camera-based traffic sign recognition can be traced back to as early as the mid-1990s. Piccioli et al. [22] described a method for traffic sign recognition using two steps: the detection of triangular and circular signs by performing a geometrical analysis of edges, and sign recognition by comparing the normalized signs (i.e. after pre-processing them to allow comparison) with image templates from a database. Although many other approaches have been proposed since then, they usually fall into one of two major categories: the (traditional) hand-crafted features approach and the deep-learning approach.

As the name implies, the hand-crafted features approach usually involves firstly a transformation of detected signs into a manually chosen feature vector (to recognize the sign's attributes) such as the hue-saturation histogram [23] or the Histogram of Oriented Gradient (HOG) [24], and secondly the training (one could say, the equipping) of a classification algorithm, such as Random Forest or the Support Vector Machine, using the output of the feature vector.

In contrast to the traditional approach, the deep-learning approach does not require any hand-crafted features. Instead, the system is trained to build its own internal representation in order to be able to output an accurate sign classification based on the training data. This is typically done by training a deep neural network – an artificial neural network that has many hidden layers – with an extremely high amount of data.

This deep-learning approach has the advantage of avoiding the difficult task of feature engineering, i.e. designing robust hand-crafted feature descriptors that optimally represent distinguishable characteristics of each traffic sign. Deep learning is currently the state-of-the-art approach for traffic sign recognition, as its accuracy rate outperforms the traditional one. However, it is generally more computationally expensive and requires a lot more training data than its traditional counterpart. Fortunately, there are several public traffic sign datasets available that were collected from various countries such as Germany [25], Sweden [26] and the United States [27].

Traffic sign recognition has become widely available in cars nowadays; thanks to strong incentives from Euro NCAP and others, it may even become standard equipment in new cars someday. Yet already now it is possible for older cars to be retrofitted with a camera-based ADAS aftermarket device in order to take advantage of speed limit information, collision warning and other ADAS features. Due to the lack of integration with the vehicle, however, these devices are limited to warning functions and cannot actively control the vehicle.

Although the performance of ADAS traffic sign recognition has improved significantly over the last decade, there are still some challenges faced by both car manufacturers and tech suppliers when it comes to deploying the technology in cars globally. For example, there can be many local (i.e. country-specific) variations on a given sign type or class, since many countries have either not adopted or not implemented the Vienna Convention on Road Signs and Signals. This convention is an international treaty signed in 1968 that standardizes road signs, traffic lights and road markings for the sake of increased road safety and ease of international road traffic [28]. More sign variations mean more data is needed to train the re-cognition system and more effort required to collect the data, as public datasets might be limited or even unavailable for some sig-nage variants. Speed limit information in the map material also needs to be updated regularly, as speed limits can change due to new regulations. Also, even though the number of recognizable sign types is increasing, there is generally still quite a limited set that can be recognized until now.

DRIVER MONITORING SYSTEM

ADAS technology has evolved rapidly over the years and with that, vehicle safety has also improved progressively. Passive ADAS components, such as airbag or whiplash protection, and active components, such as Anti-lock Brake System (ABS) or ESC, save lives and have already become standard equipment in most modern cars. In spite of all these improvements, however, car accidents with fatal or serious injuries do still happen. Sadly, a large percentage of such accidents are caused by human errors, for instance through drowsy and distracted driving. According to a traffic safety facts report published by the US National Highway Traffic Safety Administration (NHTSA), distracted driving accounted for 8% of fatal crashes and 14% of all motor vehicle traffic crashes throughout the USA in 2018, which led to 2,841 people killed and an estimated additional 400,000 people injured [29]. One of the main causes for driver distraction is the use of mobile phones, which – according to World Health Organization (WHO) – increases drivers' crash risk by a factor of four [30].

How can AI help prevent accidents caused by driver inattention or bad decisions people made behind the steering wheel? One way in which AI helps reduce distractions is through intelligent user interaction concepts in the in-vehicle infotainment systems that allow vehicle operation without requiring the driver to take their attention off the road. We will cover some examples of these in Chapter 3. Another way is to monitor the driver's state and to warn them upon signs of sleepiness and distractions, so that accidents can be prevented from happening in the first place; this is exactly the main goal of DMS. Independent organizations such as Euro NCAP already recognize the potential of such systems to promote safer roads: they include DMS in their 2025 roadmap, with new tests expected to be carried out after 2020 [31]. Now let's have a closer look at the two major driver inattention issues that DMS aims to address: distractions and sleepiness.

Distractions are defined as any activity that take drivers' attention away from the driving task [32]. The NHTSA has classified four distraction categories: visual, auditory, biomechanical and cognitive. Visual distractions refer to any distractions that lead drivers to take their eyes off the road. These include looking at a smartphone, operating buttons on the middle console, adjusting mirrors and so on. Auditory distractions are caused by noises or other auditory stimuli that hinder drivers from fully concentrating on the driving activity – such as infants crying, a phone ringing etc. Biomechanical distractions cause drivers to take one or both of their hands off the steering wheel. This is usually the case when the driver is manually operating the infotainment system, texting or eating and so on. Finally, cognitive distractions are internal distractions that turn a driver's mind from the task of driving. Some examples are daydreaming, preoccupation with competing thoughts, involvement in conversations and negative emotions (anger, sadness, anxiety). It goes without saying that these four distraction categories are not mutually exclusive. Texting, for example, is the source of visual, biomechanical and cognitive distractions, which makes it arguably one of the most dangerous things to do while driving.

Sleepiness is often described as a state of being awake but with an increased tendency to fall asleep [33]. Sometimes the terms "sleepiness" and "fatigue" are used interchangeably because they share common symptoms and are also related to each other, but they actually differ slightly in meaning. Fatigue is a subjective sensation of declining performance that occurs in any prolonged or repeated task [34]. We might feel fatigue after exercise, but not necessarily sleepiness. On the other hand, some people suffer narcolepsy or sudden sleep attack without necessarily experiencing fatigue. And whereas fatigue is counteracted by rest, sleep is the answer to sleepiness [35].

Depending on whether it is DMS' goal to detect distractions, sleepiness or both, the system needs to be trained to look for the relevant symptoms in the driver and then trigger the warning accordingly. Most visual distraction DMS employ techniques that estimate drivers' head orientation and gaze direction based on the camera image. Based on this information, the system classifies the driver's visual attention into three regions: on-road, off-road or mirrors/instrument cluster (speedometer, fuel/energy display etc.). If the driver's attention is away from road for too long or too frequently, the system will trigger a warning. The widely used threshold values are the "2/12 rule", i.e. 2 seconds for single glance period and 12 seconds for a cumulative one, as recommended by the NHTSA [36]. The two-second single glance period means that the driver ought not to look away from road longer than two seconds at a time. The cumulative period means that vehicle operations that take the driver's attention away from road, such as controlling the navigation system or the radio, can be performed over several shorter glance periods (less than two seconds each before again returning focus to the on-road region) if necessary, but the whole operation has to be completed in less than twelve seconds in total.

Gaze estimation algorithms can be categorized into two general approaches: geometry-based and appearance-based. The geometry-based, also known as the model-based approach, estimates the point of gaze using geometric calculation based on a constructed 3D model of the face or the eyes. Depending on the algorithm used, the model might be constructed using the estimated head pose usually in conjunction with the detected or estimated position of relevant facial "landmarks" such as pupil centers, eyeball centers or mouth. The appearance-based method uses machine learning to determine the point of gaze based on direct eye or face images as input. Visual descriptors, for instance, local binary patterns (LBP) or multi-scale histogram of oriented gradients (mHOG), are extracted from each image and processed through a machine learning's regressor (for numerical output) or classifier (for categorical output) algorithm. One big advantage of the appearance-based method is that it is less demanding than its geometry-based counterpart, since the appearance-based method does not require high-resolution input images to perform the gaze estimation. State-of-the-art gaze estimation employs Convolutional Neural Network (CNN) and other deep-learning methods either to replace the above (manual) feature extraction step or to act as an end-to-end system. The latter takes direct eye or face images as input, and then outputs directly the estimated gaze direction without any human pre-defined intermediate step such as the aforementioned feature extraction.

DMS for detecting sleepiness and cognitive distractions are generally based on measurements of human physiological symptoms (or signals), such as eyelid closure, eye blink rate, nodding frequency, eye movement, yawn frequency, heart rate or facial expression. Eyes are one of the most important signal cues of an individual's state of alertness. Before the eyes can be monitored, however, the system has to perform several pre-processing steps such as face detection and eye detection, which we will discuss shortly.

Face detection aims to detect whether a human face is in an image and if so, where it is. Sometimes the terms face detection and face recognition are used interchangeably, although they actually mean two different things. Whereas face detection is about detecting a face without knowing who the person is, face recognition is about identifying the person by matching the detection with a face database. Hence, face detection usually precedes face recognition.

One of the most popular algorithms for face detection is the Viola-Jones algorithm [37]. This algorithm can detect faces very fast, but performs reliably on frontal face images only. Recent deep-learning techniques in face detection, such as the family of region-based convolutional neural networks (R-CNNs), have pushed the boundary further as they are not limited to this constraint and their accuracy is typically superior to the Viola-Jones algorithm's [38].

After the face region is detected, the next step is to localize the eye regions within the face boundary. Eye detection algorithms can be categorized into four major approaches: template matching, feature-based, appearance-based and hybrid [39]. The template matching approach compares an eye template with all areas within the face image and returns the areas that best match the template. The feature-based technique looks for eye features (or characteristics) such as color and shape throughout the image. The appearance-based method uses machine learning to classify whether a certain region in the image is an eye or not. Lastly, the hybrid technique is simply a combination of all the above approaches.

As mentioned in the beginning of this section, DMS need to warn the driver upon detection of that driver's sleepiness or distraction. There are three example warning strategies as proposed in [40]. They can be summarized as continuous, critical-situation-only and "convenient" warning. The continuous warning strategy warns whenever the system detects sleepiness or distraction. The critical situation-only strategy only warns the driver in critical situations, e.g. when a collision is highly likely. The last warning strategy might not necessarily give a direct warning to the driver but rather influences the currently active ADAS functions based on the driver's state. For instance, the ACC function may allow a longer period of the takeover request time, i.e. the maximum time period before the driver is requested by the system to take back the control, when the system recognizes that the driver is fully attentive. Whereas the takeover request time might be reduced if the system detects driver distraction or sleepiness.

Although the AI realms of face detection, gaze estimation, emotion recognition etc. have been studied for years, their real-world applications as DMS in cars are still in their infancy. In order for DMS to be useful, the system has to perform reliably under many challenging real-life driving situations such as dark cabin environments, drivers wearing glasses and drivers with obscured faces. Privacy might be an additional issue, considering that most (if not all) inattention driving detection techniques use cameras as sensors and some non-safety-critical processing might need to be performed outside the vehicle (on car manufacturer's or other external servers) due to the limited resources of the on-board computing platforms.

SUMMARY

In this chapter, we looked at how AI is being increasingly employed in ADAS to support safe, efficient and comfortable driving and parking.

The field of automatic parking has been progressing from non-automatic through semi-automatic to the introduction of the first fully automatic parking systems. The latter has two levels; the less advanced one only requires a pair of eyes and the pressing of a "hold-to-run" button, whereas the more advanced version (automated valet parking) needs no human surveillance or involvement at all between the drop-off and recovery of the vehicle. We saw how increased parking infrastructure technology will continue to be needed as the parking automation level rises. Meanwhile, we considered the technical capabilities already built into today's cars in terms of radar and potentially also lidar. At all levels of automation, the vehicle's environment is constantly monitored, evaluated and responded to with the help of algorithms that are applied according to various methods and approaches.

Concerning driving on the road, we considered the context of the Euro NCAP's standards, which are used to measure the level of safety features of new cars in Europe. For navigational purposes, information is combined from a GPS, map material and automated traffic sign recognition (through camera and sometimes lidar input). We saw that using these three methods, automated traffic sign recognition faces particular challenges because of the lack of universally applied standards in traffic sign forms.

Last, we looked at driver-monitoring systems, whose purpose is to reduce the accident risk caused by drowsiness and inattentiveness. The general rule being that keeping one's eyes on the road makes for safer driving, we considered the role of visual, auditory, biomechanical and cognitive distractions (and noted how these forms of distraction can occur together – with texting while driving being a prime example). Visual attention is generally camera monitored according to the driver's head orientation and gaze direction, whereas sleepiness and cognitive distractions are perceived according to physiological signals; when a warning is needed, it will be continuous, critical-situation-only or "convenient". Challenges in the area of driver-monitoring systems include low cabin lighting and the obscuring of a driver's facial features through hairstyles, clothing or the wearing of glasses.

2

AI FOR AUTONOMOUS DRIVING

Since the beginning of the automotive industry, we have seen gradual improvements bringing increased comfort, functionality, speed and safety. The basic goal of transportation is to move people and goods from A to B, and one key aspect of that is to ensure that the people and goods reach their destination not just quickly but safely.

The auto industry has introduced better braking systems, seat belts and airbags, as well as many mechanical components that have raised the bar in vehicle safety. However, statistics show that this is not enough. Since the digital revolution, the integration of smart Electronic Control Units (ECUs) and various sensor modalities has brought vehicle safety a big step forward toward the goal of zero accidents. Recent studies have estimated that about 94% of serious car crashes are caused by human error [41], highlighting the importance of greater vehicle automation; the final target would be full vehicle autonomy, i.e. removing human error from the equation. In addition to safety, higher levels of automation allow higher productivity and reduced stress as well as better utilization of travel time. The number of vehicles on the road might also be reduced and parking areas could be repurposed.

The first ECUs were small and rather simple, and were tasked with processing small amounts of data. Take for example an early Airbag Control Unit (ACU); its main input is the measurement coming from a mechanical "G" sensor – in essence an early Inertial Measurement Unit (IMU) – which basically acts as a switch. When a sudden deceleration happens, a signal from this sensor is compared with current vehicle speed and steering angle, and corresponding actions are taken. The algorithm behind an ACU processes the data and decides to deploy or not, based on a previously hard-coded airbag deployment threshold. Typical machine-learning and data-mining methods might be required for the ACU to arrive at this threshold.

Machine learning (ML) – namely machines learning for themselves given access to data – is now commonly used in the automotive field. Its back-office uses include discovering patterns in customer satisfaction or financial trends; inside the vehicle, it is employed to monitor the drowsiness levels of the driver by analyzing face and eye movement patterns (as we have seen in the previous chapter). Applying the ML concept to more complex situations where a machine is able to perform what we would think of as a "clever" task, we refer to Artificial Intelligence (AI). This is precisely the case for Self-Driving Cars (SDC), where AI is used to enable vehicles to navigate through traffic and handle complex situations.

Autonomous Driving (AD) is without a doubt a good case study of applied AI. Many sensor modalities (including camera, radar, lidar and ultrasound) are employed to achieve AD. Data from each of these sensor modalities is processed and understood by means of ML and AI, allowing the performance of complex functions. Some good examples here are the various ECUs that allow for Automatic Emergency Breaking (AEB) or advanced Automatic Cruise Control (ACC). Other examples are the more recent smart cameras, with lane recognition capabilities that allow Lane Keep Assistance (LKA) or enable Traffic Sign Recognition (TSR).

AEB, ACC, LKA and TSR are some already-existing examples of Advanced Driver Assistance Systems (ADAS). These are only designed to help; the driver stays in charge all the time. So far ADAS have set a fairly good precedent, showing that higher automation levels can improve safety and comfort of driving at the same time. In this context, the Society of Automotive Engineering (SAE) has defined six Levels of Driving Automation, ranging from no automation (Levels 0 to 2, driver-support features) to full automation (Levels 3 to 5, automated driving features) [42].

Levels 2 and below are commonly understood as "eyes on, hands on", which means the driver is fully in the loop, paying attention and driving the whole time. Level 2 features might provide steering and brake/acceleration support, but without reducing the driver's obligation to be in control of the vehicle all the time.

Level 3 instead allows for "eyes on, hands off"; in other words, the human is not driving when Level 3 features are enabled but must still pay attention to the road and take control when ADAS features request it. Level 3 features drive only when all required conditions are met; one example is the Traffic Jam Chauffeur (TJC).

Level 4 and Level 5 refer to systems that do not require the driver to take over control; following the same analogy, one can understand them as "eyes off, hands off". The main difference between Level 5 and Level 4 systems is that the former can drive the vehicle under all conditions, while the latter is designed to drive within a specific Operational Design Domain (ODD), for example, geofenced areas as in the case of inner-city robo-taxis, or under certain lighting or weather conditions, i.e. given daylight, a clear sky or no more than a light rain.

In this chapter, we will focus on Levels 3 and above, where AI is a key enabler for SDC technology. In the following sections, we will explore AI algorithms utilized to solve the AD problem; more specifically we will look at key building blocks on modular approaches to AD: sensing, planning and acting. In a modular AD approach, a pipeline of separate components is employed, linking sensory inputs to actuator outputs. By contrast, end-to-end approaches infer control commands directly from a set of given sensor inputs; these will be further discussed in Chapter 6.

PERCEPTION

We as humans perceive our physical environment through our senses, which provide us with necessary information to assess situations we encounter. For example, if we want to cross the street, our eyes observe the scene, identify walkable ground such as pedestrian crossings and crosswalk markings, and detect other road users as well as traffic lights and other static obstacles; our ears perceive noise from approaching vehicles. All this information is put together to generate a mental map of the current state of things around us, for instance where objects are and their speed and direction. In addition, we have an understanding of our own current state within this mental map, including our speed and direction of movement. Our perception of the environment and of our own state should provide us sufficient information to plan (however quickly this may need to take place) and act accordingly.

Perceiving the environment is crucial to assessing a traffic situation. For that purpose, Autonomous Vehicles (AVs) employ various sensor modalities such as cameras, radars, lidars and ultrasonic sensors – as well as others that measure distance traveled and inertia. AVs not only employ sensors but also the right AI to understand the data they output.

Following the analogy of the pedestrian wanting to cross the road, AD systems need to detect, classify and characterize 3D information concerning road users, road boundaries, drivable space, traffic lights, traffic signs and any other element that is relevant to driving safely. In high-speed scenarios it is of vital importance to identify any hazards and threats as soon as possible; in other words, it is necessary not only to detect relevant factors in the vicinity but also at further ranges, since high speed implies greater distance traveled before the driver can intervene and longer braking distance after starting this intervention [43]. This means that sensors used in combination with AI algorithms need to provide reliable detection rates at far ranges, which translates into higher requirements for both sensor resolution and data rates.

Another important requirement for AD perception algorithms is that they need to run fast; they are an important part of a real-time system in which every millisecond counts toward ensuring safe operation. Take for example a vehicle moving at 100 km/h; this translates to 28 meters/second, which means one more second of reaction time will mean 28 meters of additional braking distance [12]. Considering that sensors need to provide enough information to fulfill required detection ranges and accuracies, and assuming that an AV system needs to understand several sensors at the same time, it is easy to conclude that a high-performance ECU is needed. Yet even if one can be employed, AI algorithms might need to find the proper tradeoff between accuracy and available computing resources. Greater performance translates to increased safety.

In view of the requirements and constraints, researchers and developers have proposed different perception methods which in turn have their particularities for each sensor modality. In addition to this, in more recent approaches several sensor modalities are sometimes combined. The majority of the most recent solutions make use of deep-learning algorithms such as CNNs. A CNN attempts to mimic biological neurons; most existing solutions belong to the so-called "supervised learning" group, in which the final CNN learns specific tasks after being taught with millions of examples. CNN usage boomed in 2012 when AlexNet was proven to outperform traditional approaches, being clearly faster thanks to the use of GPUs [44]. Some well-known and widely used CNN architectures are AlexNet, GoogleNet, ResNet and VGG-16 [45]; different sensors can require different CNN approaches. We will briefly cover some of the commonly used perception methods for cameras and lidar in the following paragraphs.

Camera-only approaches can be grouped into object based and pixel based. Object-based methods are used to detect and classify objects that could be encapsulated in a 2D bounding box within the image, for example, vehicles, cyclists, pedestrians, traffic signs and traffic lights. By contrast, pixel-based methods make use of image-segmentation techniques in order to assign a class to each pixel, which is particularly useful for those elements that cannot be encapsulated within a bounding box, for example, a road surface or lane markings.

Object-based methods can be further categorized between so-called "single-shot" algorithms and region-based algorithms. Examples of single-shot algorithms are You Only Look Once (YOLO) [46] and Single Shot Detector (SSD) [47], which perform both detection and classification within the same Deep Neural Network (DNN). First, the image is divided into cells, forming a grid. Then features are extracted and class probability is computed for each grid cell. Region-based algorithms, also called two-stage methods, first detect Regions of Interest (ROI) and then apply classification to them. For example, circular shapes within the image might be detected as ROI for TSR. Once a ROI is detected, a classifier will tell if it is a real traffic sign and might even determine what traffic sign it is. Region-based methods produce better object recognition and better object localization within the image, while single-shot methods are faster and therefore more commonly used in real-time applications [48].

Image-segmentation methods classify each pixel of an image into a specific class – for example, road, vehicle, pedestrian, tree, building or sky. Image segmentation has higher computational requirements, which until recently made it unfeasible for real-time applications. These methods can be further categorized into those that use Fully Connected Networks on a single frame, for example,

DeepLab [49] – which is state of the art in terms of accuracy but also computationally expensive – or else Enet [50], which is proven to offer a good trade-off between accuracy and efficiency [51]; and those that utilize temporal information, such as Long/Short-Term Memory (LSTM) approaches [52], [53]. The first group might also combine a priori knowledge of structures, for example, the likelihood of certain parts of the image being sky – or road or other structures – which helps to optimize precision and run-time.

Mono cameras tend to fall short in producing 3D information. Still, stereo camera systems can produce 3D information, although 3D reconstruction accuracy is constrained by image resolution, baseline length between the stereo-pair (i.e. the distance between the two cameras which produce two slightly differing images of the same thing), and the distance of the 3D object. By contrast, accurate 3D information is the main contribution of lidar. Lidar's basic principle is to measure the time it takes for a light beam to travel from its source, bounce back from a surface and return to the source. Modern lidar "broadcasting" sensors can transmit hundreds of thousands of such light beams per second into predefined directions; their reflections produce a cloud of 3D points that indicate spatial position. Since it is assumed that each 3D point belongs to a certain object in the scene, after identification each point will also have to be "clustered": assigned to the correct object.

Lidar point-cloud processing approaches could be grouped into two main categories: the ones that transform 3D point clouds into 2D data and use 2D image techniques, and the ones that work directly with 3D point clouds to identify shapes. Examples in the first group either project lidar 3D points onto an image plane and cluster them based on image object detection (techniques for detecting objects within images) – or else map a 3D point cloud into a 2D top view or into a 2D depth map, and then apply 2D image classification. By contrast, the second group avoids usage of 2D planar images; instead, 3D data is directly analyzed, either by compartmentalizing it into a grid of cells – often referred to as Voxels (using an algorithm such as VoxNet, SparseConv or 3DMV), or by running

CNNs directly on 3D points (employing e.g. the PointNet or SPLATNet classification network). Such methods are used to segment (i.e. divide up) the ground plane (i.e. the road surface) [54], to detect 3D vehicles [55] or to better understand the exact shape of curves and curbstones [56].

We have seen methods for object detection and classification in the context of perception. We should notice that detection and classification are two important building blocks, but perception extends beyond that: perceiving the environment also includes the use of relevant algorithms for object tracking and multi-sensor data fusion, among other things, to allow proper understanding of the surrounding conditions. Although detailed explanation is beyond the scope of this book, we hope to have given you a good introduction to the subject here.

PLANNING

Once the vehicle has identified nearby road users and "understood" the road boundaries and drivable path ahead, it is time to plan the best action. Planning is a complex task that involves different layers: global mission planning to find the best route from A to B following the existing traffic rules; behavior "planning", which involves estimating the most likely behavior of all external actors in the scene; and local planning to generate a continuous trajectory – which means the planning of immediate actions to bring the vehicle to next desired state (position).

Imagine the situation where we want to cross the street. Our goal is to reach the bakery which is at the other side of the street a few meters ahead; this is our global plan. We have now perceived the scene, we have chosen the best spot to cross that street, we understand what space is walkable, we have estimated the other actors' current state and we guess their most likely next state – in other words, we "plan" their behavior – and so we are ready to plan our next move. We have multiple possibilities that could bring us to reach our target destination. We could continue walking on the same side of the street, approach the closest crosswalk or attempt to cross right away etc. To decide which of these options are doable, which is safest or which is optimal, we need to evaluate all given information and plan accordingly.

The analogy also applies to AVs: having perceived its surroundings, the vehicle needs to contemplate multiple hypotheses and choose the one most suitable to bring the car to its final destination – as safely and quickly as possible and complying with traffic rules. To achieve this goal, the vehicle needs to localize itself within a given map so that the best route can be calculated; at the same time, it needs to predict likely behaviors from other road users and finally to compute the most immediate trajectory to follow.

On a global scale, the AVs "wants" to reach a destination from the known current position, which we can assume it is given by the combination of on-board global navigation satellite systems and sophisticated localization algorithms that are used to localize and associate static objects with previously mapped objects such as lane markings, poles and traffic signs. The best route is then calculated, given different parameters such as traffic rules, a priori traffic knowledge, current traffic conditions, distance etc. Traffic rules, street distances, and all other relevant information about the streets concerned are part of existing maps. Maps are typically represented in the form of graphs, while current traffic density and other real-time information are available through backend server communication including Vehicle-to-Vehicle (V2V) and Vehicle-to-Infrastructure (V2I). A graph consists of nodes and edges connecting these nodes; one can imagine a road intersection as a node and the roads reaching it as the edges. Each of the edges might be connected to another road intersection node at its other end. One can intuitively imagine each of these edges to be associated with a certain distance, maximum allowed speed, direction, current traffic information etc., which in graph theory is referred to as its weight.

In order to plan a route, one needs to traverse the graph (basically this means to evaluate its information) to find the best path that fulfills the desired goal; the goal might be the fastest path, the shortest one in terms of distance or maybe even the one with the least traffic lights. To do so, several graph-traverse algorithms have been widely used, among them the so-called Best-First Search which consists of moving through the graph by following the most promising nodes given certain predefined rules. As good examples of best-first algorithms, Dijkstra and A* both look for the path that minimizes a certain weight or cost (such as time, fuel cost or distance) between two nodes (which represent current position and desired destination). Of these two algorithms, A* is more accurate and efficient. In order to make these methods real-time capable on large-scale maps, incremental graphs are usually employed. Best-first methods like A* rely on certain predefined heuristics, whereas some more recent approaches make use of AI in an attempt to optimize the selection of such rules or even look into new ways to approach route planning and logistics by employing Recurrent Neural Networks (RNNs). The latter can look into temporal sequences (for example, a series of sequential vehicle position estimates over a certain length of time) that are well suited for path-planning applications [57].

In terms of behavior planning, there is extensive research analyzing other road users' possible behavior, for example in order to understand and predict whether an oncoming vehicle will continue straight or maybe turn and interfere with our path. Human driving behavior can be learned by imitating (observing and repeating) real human driving experience data. However, there are corner cases (cases involving a problem or situation that only occurs outside of normal operating parameters) such as crashes, dangerous situations and unlikely events that either cannot be recorded or of which the number of logged examples is limited – with the result that these behaviors cannot be learned from real data. Instead, simulated data can be used to train (i.e. rehearse or gain knowledge of) such situations, learning by means of Deep Reinforcement Learning (DRL). DRL is the combination of two AI techniques: reinforcement learning and deep learning. Reinforcement learning algorithms are based on combining and utilizing the positive results from sequential trial and error, while repressing the negative ones, in order to decide which action would be best. Deep learning is simply a more popular term to describe ML based on neural networks with many hidden layers that enable the evaluation of complex input data.

To plan the best trajectory, several implementations are based on "fitting a curve" to (i.e. constructing a curve representation for) a set of possible-way points; this may, for example, be defined as the middle path between the lane markings that are perceived by a front camera. Fitting a curve can be done using several different curve construction models (for instance polynomials, splines, clothoids or Bézier) that are useful for generating smooth trajectories. Other implementations use random sampling or optimization of the possible navigation area. The so-called Rapidly exploring Random Tree (RRT) [58] is an example of a random-sampling trajectory-planning algorithm; these algorithms cannot guarantee optimal trajectories. Optimization-based methods can produce optimal trajectories, but their higher computational needs can present critical impediments to real-time operation.

The current trend in automotive research focuses on learning-based approaches, as used for example to plan optimal lane-change trajectory [59]. Another example of AI-based trajectory planning learns spatial-temporal features on a neural network [60] that is LSTM based.

MOTION CONTROL

Of course, the next step is to move based on what has been planned – i.e. to convert the plan into executable actions. Going back to the pedestrian going to the bakery, if we assume that there is a crosswalk ahead – while at the same time there are vehicles coming from both directions – the plan might be to stop walking until there is no more incoming traffic. This means that our body is told to stop. The action "stop" can be smooth, sudden, fast, slow, continuous, interrupted etc.; the chosen dynamic means applying more or less energy for an appropriate amount of time in order to move in a controlled manner. Humans do not tend to think about the implications of such actions, but it is actually a complex task to send the right signals to the multiple muscles that are our actuators.

The same principle applies for AV; the right signals must be sent to various actuators such as the steering and braking systems, engine, transmission and suspension. These signals are not a discrete one or zero – not an "all or nothing" – and it is the task of the motion controller to map the intended action into signals that actuators can understand – a for example, the amount of pressure a braking system needs to apply to reach a full stop in a certain given time.

Motion controllers rely on real-time knowledge of vehicle state information, for example concerning speed and orientation, for which it is necessary to estimate the right vehicle dynamics model. One can divide the techniques for estimating vehicle dynamics models into two main groups: model based and data-driven-based. Model-based methods are generally variations of the well-known Kalman Filter (KF) or Particle Filter (PF), which optimize the estimates of the vehicle when there is measurement error but depend on the reference vehicle model used. On the other hand, data-driven estimation approaches avoid model-based limitations by estimating vehicle state and parameters through learning from historical and real-time data. This learning process is done through AI with the use of neural networks [61].

With the proper vehicle dynamics estimated, motion controllers can now send the commands that will keep the vehicle as close as possible to its target trajectory. Here one can consider traditional solutions such as Proportional-Integral-Derivative (PID) controllers; these are widely used in Active Cruise Control (ACC) systems. PID controllers employ a feedback loop, continuously calculating the difference between desired and measured, and applying corresponding corrections. PID controllers depend on a set of fixed parameters, however, which limits their use to a set of predefined use cases. There is a growing trend toward employing "learning controllers", which make use of training data. Learning controllers are capable of anticipating repeating vehicle motion effects and disturbances, thus improving safety and comfort [62]. These learning controllers use CNNs, for example, the so-called Iterative Learning Control for path tracking in SDC, or Model Predictive Control which computes control actions by solving an optimization problem. DRL and imitation learning have been employed to derive discrete control actions, for example, turn left, turn right, accelerate, brake [61].

SUMMARY

The first wave of AI has been shaping AD research and development, it is already part of existing ADAS solutions and it will gradually be introduced into more vehicles – especially given the trend towards software-defined cars whose AI algorithms can be improved and deployed with Over the Air (OTA) updates. We can already observe many prototypes and production vehicles on public roads that are part of the early learning phase for AI-based SDC.

Simple use cases can be addressed with classical computer vision (i.e. trying to make computers visually perceive like humans do), path planning and motion control algorithms, but it is thanks to AI that we are starting to solve corner cases and more complex scenarios where traditional algorithms would usually fail.

If we simplify the complex problem of AD into three main blocks – sense, plan, act – we have seen in this chapter how AI already plays an important role in sensing, with so many pattern-recognition perception algorithms available. We have also seen that planning and motion control can strongly benefit from the use of AI. Existing methods use more traditional approaches, which in general require handcrafted parameters or predefined rules – which can be automated and optimized using AI and ML techniques. Moreover, we have seen that research and development is moving towards AI- and ML-based approaches; some of these will become the status quo in next-generation AD vehicles.

There is a growing trend among new AI algorithms to be deep learning based. Convolutional Neural Networks (CNNs) are commonly used to process spatial information, for example for object perception in still images/individual frames. RNNs are used to process temporal data, for example, video streams; this method is particularly well suited for path-planning algorithms, which is where LSTM algorithms can come into play. DRL algorithms are commonly used in AD, for example, to learn driving optimal driving policies to bring the car to its destination.

All this is just the beginning. Moving forward, AD will gradually become reality for mass production, and this will certainly happen in association with innovation and new developments in AI. In other words, safe and robust AD will be associated hand in hand with AI.

3

AI FOR IN-VEHICLE
INFOTAINMENT SYSTEMS

Over the last two centuries, the automobile has gone a long way from a mere means of transportation (you could say a "horse replacement") to a kind of motorized personal connectivity device. But it was not until the beginning of the 20th century that cars started to gradually lose their status as the exclusive property of the rich and privileged. Thanks to Henry Ford, who revolutionized the mass production of automobiles in the late 1910s, cars became affordable for many people and the concept of cars as widespread personal transportation began to emerge.

As car manufacturers constantly look for ways to stay ahead of competition, customers' growing expectations can no longer be easily satisfied by a product solely capable of transporting them from A to B safely. The whole driving experience has become the new Unique Selling Point (USP). Hence the in-vehicle information and entertainment system, or infotainment system for short, plays a significant role in making the time spent inside the vehicle pleasurable for both driver and passengers. It is unthinkable nowadays for a car manufacturer to make cars for the mass market without at least some basic infotainment-related features, such as a radio, preinstalled.

Before going into more detail about how AI can be applied in infotainment systems, let us examine more closely what these systems are. As the name implies, infotainment systems are simply a general term referring to a set of hardware and software that provides information and entertainment for the vehicle's occupants. They range from built-in navigation systems, rear-seat entertainment screens, Bluetooth speakerphones to audio/video media players and many other features. A special software component called the Human-Machine Interface (HMI), which usually runs as part of the infotainment system, coordinates the display of information and controls the system's functionality (and sometimes also other electronic systems in the car, such as the air conditioning) based on user interaction over various in-vehicle control elements. Information may be presented simultaneously using various displays, e.g. the middle console, the cluster instrument display or the head-up display. User controls may take many forms, ranging from simple buttons on the steering wheel, touchscreen or touchpad, to touch-less technology such as voice command and hand gesture control, which we will discuss briefly in the following sections.

The rapid innovations coming from the world of consumer electronics – in fact, from smartphones – undoubtedly pose a major challenge to infotainment systems' designers. For example, the average age of a passenger car in the E.U. is around 11 years [63]; it's no surprise that not all infotainment systems can keep up with the latest technology using software and hardware that may have been designed at least a decade earlier. Furthermore, as infotainment applications become increasingly sophisticated or computationally intensive or simply require amounts of data that might not fit into the car's hardware-constrained embedded systems, they become more dependent on external (extra-vehicular) computing resources such as cloud servers. However, the longer the car is "online", the larger its "(cyber-)attack" surface is. In other words, "connected" cars are generally able to offer better service and/or overall user experience than their "offline" counterparts, but that comes at the price of higher security risks. Infotainment systems usually act as the "Internet gateway" for the rest of the vehicle, basically making them the component most vulnerable to security breaches.

In the following sections, we will briefly cover three examples of infotainment system applications that are empowered by AI.

GESTURE CONTROL

Gestures are a way to convey a specific message non-verbally by moving one's hand, head, finger or another body part in a certain way. In contrast to random actions, only meaningful or intentional movements can be considered gestures. We use gestures in our daily lives, even without thinking about it. For example, we may point our thumb upward to show approval or downward for disapproval. Gestures can also be used when we need to communicate with someone who does not share a common language with us.

The idea of controlling computers through the use of gestures is not entirely new. Research into gesture control for human-to-computer interaction can be traced back as far as 1980 [64]. Since then, popular science-fiction movies featuring gesture control in futuristic human-to-computer interaction, such as "Minority Report" (2002) and "Iron Man" (2008), made the concept of gesture control as an alternative method for interaction with computers more widely known.

The use of gesture control as one of the user input methods in cars promises several benefits, including improved user experience and better safety through less distraction. Pointing toward the latter, a study conducted in the Netherlands by van Nimwegen and

Schuurman of Utrecht University showed that drivers spend significantly more time looking at the road when they are able to perform secondary (non-driving) tasks, such as controlling the air conditioning system, using gestures [65].

Although gesture recognition and control have been actively researched for over four decades, it was not until recently that the technology was adopted in production cars. The 2016 BMW 7 Series premiered the usage of gesture control as a means of touchless interaction with their infotainment systems [66]. Since then gesture control has also become available from other car manufacturers, supporting at least some basic hand and finger gestures to e.g. adjust sound system volume or change playlist tracks.

Gesture recognition systems typically use a camera that is mounted on the vehicle's ceiling and that faces down to the area in front of the middle console. For the recognition to work reliably regardless of ambient light conditions, camera-based systems are usually equipped with an additional filter that removes the light spectrum not relevant for the recognition process (e.g. a daylight filter); at the same time, they also typically add some illumination, for instance from near-infrared light-emitting diodes (NIR LEDs) [67]. Although invisible to the human eye, light emitted by the NIR LEDs significantly aids recognition by enhancing the contrast between the foreground object and its background.

An alternative approach is to use radar instead of cameras [68]. Unlike camera technology, radar works dependably regardless of ambient light; it is also more "user-privacy friendly". On the other hand, radar-based systems usually have lower resolution than their camera-based counterparts. For example, they may not be able to distinguish small gestures originating from moving body parts located very close to each other.

Another system that has been tested takes the best of both worlds by combining radar-based and camera-based sensors [69]. Through the combination of short-range radar, a color camera and a time-of-flight (TOF) camera, overall gesture-control system accuracy and performance in varying operating conditions can be considerably improved while significantly reducing power consumption. Although there are considerable advantages with this system, it tends to be the most expensive configuration since it requires more sensors.

Now let's examine how gesture control works. A gesture-recognition system based on cameras or other input sensors as mentioned above is at the core of this technology. Yet since most gesture-control systems in production cars as of this writing (2020) are dependent on cameras, we will limit our discussion here to camera-based systems.

Although the technical terms might differ from one system to another, the gesture-recognition process is typically comprised of three main steps: detection, tracking and recognition [70].

Gesture detection is the first step. Relevant objects, e.g. hand or fingers, are localized and distinguished from non-relevant ones (regarded as background) in an image. This task is sometimes also referred to as segmentation, as it partitions the total image into areas of relevant objects and background ones by assigning every pixel in the image to one of those areas. To improve detection, the raw image frame acquired from the camera may undergo some filtering processes such as "noise" reduction (reducing glare etc.) and contrast enhancement.

After the relevant objects are identified, the system tracks their movements in successive image frames. Tracking is essential in almost any object-recognition system. After initial detection, tracking enables estimation of the object's location in subsequent frames even though the object itself is temporarily undetectable. In the context of gesture control, tracking also supplies the information of an object's trajectory over time, which is a prerequisite for dynamic gestures: gestures that include subsequent movement over time, for example, a left-to-right hand swipe. By contrast, static gestures such as the victory (V) sign or a thumbs-up do not have any temporal attributes; they can be directly interpreted without having to first observe their trajectory.

Next, the recognition step classifies the results from the previous steps into a gesture within the pre-defined set of supported gestures using a machine-learning classification algorithm. The classifying algorithm, such as a Support Vector Machine (SVM) or Convolutional Neural Network (CNN), outputs the "most probable" gesture: the gesture that has the highest confidence value relative to other gestures in the set. Finally, the identified gesture triggers the HMI to perform the intended action, e.g. to increase the audio volume.

VOICE ASSISTANT

Thanks to the abundance of free information on the Internet and the power of search engines, it has never been easier to get and use information. We can learn instantly whether or not it will rain in our area tomorrow. Or we can look up a foreign word and within seconds, be told what the word means, how to pronounce it and many other things about it. Thanks to Apple Siri, Amazon Alexa and other services, even typing in our question or search query to a computer or smartphone is no longer necessary. We can literally talk to our smartphone, smart speaker or other smart device that supports voice assistant. Combining these input possibilities with smart home connectivity, we can also "voice command" our smart devices to control other electronic appliances at home, for instance, to switch off the light in the kitchen or to lower the sun blinds in the living room.

Voice control has long been embraced within the offering of multimodal user interfaces in cars as well – and for good reason. Not only does voice control allow better comfort for drivers; it saves lives too. According to a report published by the U.S. National Highway Traffic Safety Administration (NHTSA) in 2019, distraction accounted for 8% of all fatal crashes occurring on U.S. roadways in 2017 [29]. Hands-free user interfaces such as voice control improve safety by helping drivers keep their eyes on the road.

In the mid-2000s, Honda and IBM together introduced the first in-vehicle navigation system with speech recognition and text-to-speech (also known as speech synthesis) capabilities in production cars [71]. The system recognized spoken street and cities names throughout the continental United States and provided turn-by-turn navigation guidance through its speech synthesis technology. Nowadays, infotainment systems with voice capabilities have become more common and are offered by all major manufacturers.

Speech recognition and synthesis are two important examples of how pervasive and influential AI-based technology has become in our modern lives. In AI research literature they represent the two most prominent sub-fields within speech-processing technology. Yet although other subfields such as speech coding, speaker recognition and speech enhancement are less well known, their applications may be examples of AI technology we have used without knowing it. Speech enhancement technology's purpose is to make speech more understandable, e.g. when used in hearing aids, by automatically reducing unwanted background noise. And advancements in speech coding technology now allow us to enjoy good quality Internet phone calls without using much bandwidth.

Car manufacturers use two general strategies to integrate in-vehicle voice assistant technology into their products. The first is to employ third-party embedded voice assistant software, from Cerence (formerly known as Nuance) or another manufacturer, as a fully integrated component of their infotainment systems. Most if not all speech processing tasks (speech recognition and text-to-speech) are performed by the on-board computer and typically do not require an Internet connection. This strategy is sometimes called the "white-label product" strategy, because the integrated voice software becomes an indistinguishable part of the overall infotainment system; the user doesn't know the origin of the individual software components being used.

The other strategy is to use online voice assistant platforms such as Amazon Alexa or Google Assistant. In contrast to the first approach, here most speech processing tasks are performed online in cloud servers. In order for such online services to allow users to control aspects of the vehicle, e.g. setting the air conditioner's temperature, the car manufacturers need to support the service's application programming interface (API) and provide some access to the in-vehicle network. Because the complex tasks of speech processing are performed at powerful server farms with virtually unlimited resources, the quality and capability of these online-based systems are far superior to the (offline) onboard-only solutions that run on resource-constrained embedded systems. However, the downside is high dependency on Internet access. Without Internet connectivity, the service can either become completely unavailable or operate with limitations – since it is forced to rely entirely on its "offline" knowledge. In the latter case, the system might only be capable of recognizing a limited vocabulary set or some basic vehicle control commands.

Although the use of voice assistant technology in vehicles is becoming increasingly popular, there are many cars (especially older or less expensive ones) that do not have built-in support for it. Yet the use of an online voice assistant service is actually still possible in these cars if the car is connected to either a smartphone or an aftermarket device. However, since these approaches are not integrated into the vehicle, they cannot be used as a hands-free user-interface alternative to operate controls or query the car's internal information – unless they have access to the vehicle's bus, e.g. through the on-board diagnostics (OBD-II) port. But non-integrated approaches do enable users to use voice technology to access information from the Internet and to use connected online services or smart devices. The latter is usually made possible with the help of various third-party extensions known as "skills" or "actions". The services or products offered by these extensions enhance the core capability of the voice assistant service, allowing users to perform custom tasks.

At the heart of every voice assistant system lies its automatic speech recognition (ASR) and speech synthesis technology. In this book, we will stick to some basics of ASR, since it would take a separate volume just to explain the complex processes that make up these two technologies.

ASR's first step is to convert the continuous (analog) sound wave recorded by the microphone into a sequence of individual, point-in-time samples – making it possible to digitally register them. These samples represent changes of sound intensity (also known as amplitude) over time and are also called the speech waveform. After the analog-to-digital conversion, the speech waveform undergoes an acoustic analysis process. This process basically slices the waveform into small "chunks" called frames and extracts the acoustic features of each frame. The most relevant acoustic feature is described by the Mel Frequency Cepstral Coefficients (MFCC). A detailed explanation of how MFCC are extracted is beyond the scope of this book; it's enough to point out here that every sound – or in this case, phoneme – has unique MFCC and these coefficients can be used to distinguish one sound from another. (A phoneme is the smallest unit of sound that together with other such units form a word. For example, the word "when" is made up of the three phonemes /wh/, /e/, and /n/, whereas the word "hen" has three slightly different phonemes: /h/, /e/, /n/.)

After the acoustic features are extracted, the most likely sequence of phonemes is calculated with the help of an acoustic model extracted from training data. Following one popular acoustic model called GMM-HMM, Gaussian Mixture Model states are used in the Hidden Markov Model system. In this approach, GMM creates a model of the probability of acoustic features to be given to a phoneme by using a mixture of several Gaussian distributions. (Gaussian distributions are also known as normal distributions; they are essentially a series of bell-curve charts plotting "states" – which we can think of as acoustic snapshots in this case – in order to predict likelihood.) The HMM system then models the probability of one particular phoneme being followed by another one.

Besides acoustic models, common speech recognition systems also use pronunciation and language models as inputs. A pronunciation model (also known as a lexicon in other research literature) is simply a dictionary that maps words to their sequence of phonemes in a given language. In other words, a pronunciation model contains a list of (recognizable) words and the rules concerning their pronunciation. This model is typically defined by linguistics experts for a specific language and should ideally support different varieties of pronunciation within a language as well, such as American and British English.

Language models, on the other hand, represent the occurrence probability of a sequence of words in text. Unlike the human-written pronunciation models, language models are trained from a large set of texts or text corpuses. One example of the latter is the Google Books Ngram Corpus, which contains data from over 8 million books – representing 6% of all books ever published [72]. Language models allow the system to favor more frequently used words or phrases over less frequent ones. This is especially helpful when they share the same sounds, for instance in the questions "But can you hear?" and "But can you – here?"

Finally, the acoustic, pronunciation and language models are merged into one huge HMM, in order to determine the most likely sequence of words to have generated the speech input. This process, also known as decoding, is basically finding the path throughout the HMM whose total probability is the highest. Due to the sheer number of possible paths, a naïve approach enumerating all imaginable paths in such a huge model would simply be too computationally "expensive" – too unimaginably large a job. Instead, a more efficient decoding algorithm such as the Viterbi algorithm [73] is usually used.

The above automatic speech recognition approach is known as the statistical ASR and has been the dominant method for several decades. Starting around the year 2010, however, new approaches employing deep learning in one way or another have been seen as state-of-the-art, as they outperform the conventional GMM-HMM approach. There are basically two paradigms applied to the use of deep learning in ASR: the hybrid and end-to-end approaches.

In the hybrid approach, deep learning is used to replace – or in combination with – one of the steps in the conventional (statistical) process. For instance, the DNN-HMM hybrid approach uses deep neural networks (DNN) instead of GMM for phoneme recognition as part of the HMM-based acoustic modeling mentioned above.

By contrast, the end-to-end approach uses deep learning for the whole recognition process. Here the speech input waveform is converted directly to words without any intermediate steps such as feature extraction or acoustic modeling. The end-to-end ASR process pipeline is much simpler than the conventional and hybrid approaches and requires no domain expertise to design. However, this approach requires significantly more training data and computing resources than the other approaches.

USER ACTION PREDICTION

Predicting future actions based on patterns learned from historical data is undoubtedly one of the most common AI applications. Although human behavior is generally unpredictable, we still tend to exhibit certain behavioral patterns or perform repetitive tasks in our daily lives. These could include the time we leave our house to go to work every day, the route we take to commute, the usual time of day when we listen to our favorite radio or podcast station and many other recurring events. Wouldn't it be nice if AI could recognize all these patterns automatically for us and proactively help us perform some of these repetitive tasks at the right times and under the right circumstances? In fact, this scenario might soon become reality, thanks to AI-empowered infotainment systems.

Recently, the German car manufacturer Daimler introduced a user-action prediction feature in their Mercedes-Benz infotainment systems (MBUX) that can produce personalized suggestions learned from past user actions. For instance, if the driver regularly places a call to someone at a certain time of the day, the phone number will be suggested when that time comes. Or if a user listens regularly to a particular radio station at a certain time, that radio station will be suggested on the display at that time [74].

In the user action prediction scenarios above, the AI system learns continuously and improves its prediction based on the history of user actions over time. The more data available for the learning process, the better the prediction will generally be. Closely related to the subject of user action prediction is user personalization. Because the system is trained using accumulated historical records of events and actions from a particular user, the resulting prediction or suggestion is also personalized to that user.

We're also seeing AI-based predictions become increasingly ubiquitous in our daily lives outside the car domain. When shopping online, recommendations appear of items that might be of interest to us based on our purchasing history. We receive personalized news feeds from our online news portal or social media apps, based on our past browsing activities or personal preferences. Our favorite online movie subscriptions can also suggest other films that we might be interested in, based on our viewing history.

All this is possible thanks to recommender systems. Sometimes these are also referred to as recommendation systems or engines. Recommender systems are a subfield of AI aiming to give users relevant recommendations of products, actions etc., based on their purchase history, past decisions or preferences and other information.

Depending on the implementation approach used, recommender systems are generally divided into four major categories: collaborative filtering, content-based filtering, knowledge-based and hybrid method. We will briefly examine each of these methods in the following section.

The collaborative filtering approach generates recommendations based on the accumulated information from other users who share similar profiles and preferences. If a number of similar users rate a specific item highly, it is possible that the current user would find the same item similarly interesting or relevant.

Note that the recommendation generated by the collaborative filtering approach works without any knowledge about the item itself (except for its rating). By contrast, the content-based filtering approach identifies the intrinsic attributes or common characteristics of items preferred by the user; then it generates recommendations of other items that share similar characteristics. So for example, if a user likes science-fiction movies and we know that a "Star Wars" film belongs to that category, it's probable that the user will like that film as well.

Both the collaborative and content-based filtering approaches rely on sufficiently large amounts of data in order to make useful inferences possible. In the cases where a new user or item is introduced, these systems perform poorly until enough data becomes available. This is known as the "ramp-up" or "cold-start" problem. By contrast, the knowledge-based approach is based on the explicit requirements or constraints specified by the current user; the system generates recommendations that match the criteria. Simply put, regardless of how positively other users with similar preferences rate the items (the collaborative filtering method) or how likely it is that the items belong to the user's preferred choices (the content-based approach), the user will not find any recommended item costing more than $100 useful if their budget is a maximum of $100. As good as this approach may seem, however, it is not without drawbacks. Knowledge-based recommender systems typically suffer from a problem known as the "knowledge acquisition bottleneck". A simplified explanation of this problem is that the performance of the system depends highly on how well the knowledge is acquired (from the subject-matter experts) and represented in the system's knowledge base; the process to accomplish this is a long, difficult and error-prone one [75]. For instance, an e-commerce recommender system would need to have internal knowledge about what products belong to the "cars" category, including the relationship between car models and manufacturers, and the properties of each car model (how many doors, body type and so on), in order to become "intelligent" enough to recommend a Porsche 911 instead of a Volkswagen Golf when the user is looking for a sports car.

The hybrid approach is simply a method that combines two or more of these approaches, aiming to both profit from the strengths and overcome the shortcomings of different individual approaches. Some common "hybridization" strategies include weighted, cascade and feature augmentation. In the weighted strategy, the score of a recommended item is calculated as the weighted sum of all individual recommender systems' scores. In the cascade strategy, the recommender systems are executed one after another in a sequence that is prioritized according to factors that include probability – but especially also suitability to the use-case; the output of the higher priority system is then refined by the next (lower priority) one in the pipeline [76]. The feature augmentation strategy is similar to the cascade method, in that two or more recommender systems are executed in a sequence. However, in feature augmentation the output of the previous recommender is also used as an input of the next one – which is not the case in the cascade approach.

In order for the user action predictions in the infotainment system domain to be useful, the recommender system deployed also needs to be context aware. In spoken or written language, context (which we can generally define as the surrounding words) helps us understand the meaning of a word better. In the case of recommender systems, context is any affiliated information including time, location and even weather etc. – that might be relevant for making better recommendations for the user. The automatic phone number suggestion use-case mentioned at the beginning of this section is one example of these context-aware recommender systems in action. However, adding context increases both the dimensionality and sparsity of the system's model [77]. In other words, since the system becomes significantly bigger and more complex, training it to perform well requires a lot more data to "fill it in".

As in the case of all AI subfields, deep learning also plays an increasingly significant role in improving recommender systems. From the image recognition domain, we have seen that deep learning can capture the essence of an object from input data much more effectively than human-engineered features (i.e. attributes of that object defined by humans and put manually into the system) can. This makes deep learning a promising approach to bring the quality of recommender systems to the next level – as it may even be able to effectively identify the underlying relationship between user and item considering all available information about context, thus leading to a better recommendation or user action prediction.

Many types of recommender systems based on deep learning have been investigated in the literature, ranging from multilayer perceptron and CNN to attentional models and Deep Reinforcement Learning. A full discussion of these approaches goes beyond the scope of this book, but interested readers may wish to refer to the survey of these systems made by Zhang et al. [78].

User action predictions and other AI-based personal recommendations will increase comfort, usability and the overall user experience in automotive applications when the user finds them useful and relevant. Taking this view further, they can also bring safety improvements as they have the potential to reduce distractions by filtering out irrelevant information and requiring less interaction from the driver. However, the opposite result is also possible: poor recommendations or predictions could be more annoying than helpful or could lead to unnecessary distractions.

Another potential issue arises from the fact that recommender systems learn by observing a user's behavior and actions in many contexts. This might raise privacy concerns as the system can only improve its predictions over time by collecting data regarding personal preferences, times of activity, history of locations and other sensitive information. Depending on the algorithm, the amount of data, the on-board computing power and other considerations, the collected personal data may need to "leave" the vehicle in order to be efficiently processed (and potentially stored) on the car manufacturer's back-end servers. At the same time, the level of data protection might vary from one car manufacturer to another due to differing security architecture, IT infrastructure, cybersecurity capabilities, data anonymization/pseudonymization concepts and many other factors.

SUMMARY

In this chapter, we have considered how important AI has become in the evolutionary process that has taken automobiles from eccentric and expensive horseless carriages to commonly available consumer items outfitted with a range of technically sophisticated support systems. The fact that many of these functions are linked to the car's in-vehicle information and entertainment system points to the growing importance of creating a pleasant driving experience as a whole.

We saw how the convenience of AI-driven gesture control can also add to driving safety. The gesture-control process starts with detection through the use of cameras, NIR-LEDs and/or radar and is followed by tracking (which can fill in the blanks if the object temporarily becomes undetectable) and recognition (through calculating the probability of the gesture's meaning) before cueing the HMI to perform the desired task.

We also saw how voice control similarly improves both comfort and road safety by keeping drivers' eyes on the road. We considered a few features of the speech-processing technology that enables voice control, especially looking at speech recognition and synthesis. Last, we had a look at how AI is helping us to not forget regularly recurring tasks by recognizing patterns in our actions (such as making specific phone calls or listening to a radio station at consistent times). The recommender systems employed to do this use filtering processes that are either people based (the collaborative approach, comparing us to other users with similar profiles and preferences), general preference oriented (the content-based approach; what does the user like?), oriented around stated customer requirements (the knowledge-based approach) or some hybrid combination of these three processes. Context awareness also plays into the AI equation here, as does deep learning. Privacy concerns may enter in as well – especially when data concerning personal habits are processed on external servers.

4

AI FOR RESEARCH & DEVELOPMENT

Traditionally, the automotive industry brings the value to customers at the point of sale of a new product. The newly sold vehicle is at its peak of technology and value on the day the customer receives it; from that moment on, the value of the car diminishes. One of the reasons for this decrease is the fact that technology will continue advancing and evolving, making parts of that vehicle less valuable or even obsolete. Until recent years, a car manufacturer would develop a new feature, for example, AEB, LKA or ACC, and would perform a series of verification and validation tasks to ensure that it dependably performs at its peak when the vehicle leaves the factory. This approach implies that these functions are at a fixed level of development and will not improve any further. However, the current trend in the auto industry is to include the possibility to update over the air (OTA) [79], similar to how computers and cell phones allow for operative system and applications upgrades to latest versions with new functionality and features and improved security; this also increases customer satisfaction. This kind of approach not only allows the manufacturer to upgrade functionality but also to fix or solve software issues without needing to physically recall vehicles – producing enormous savings, since faulty software might concern millions of cars.

This is what we call a software-defined approach [80]: one in which software is at the core of the technology. This implies that this SW (software) is running on top of a sufficiently capable platform – in other words, is backed by a computer that can allocate new functions and upgrades over time. Such an approach is key to holding value long after the point of sale, and its value essentially comes through the use of AI.

We have seen in previous chapters how AI is revolutionizing Advanced Driver Assistance Systems (ADAS), Autonomous Driving (AD) and in-cabin experience by using Intelligent Cockpit and infotainment functionalities, for example voice assistant. The automotive industry is not only benefiting from AI within the car itself, but also in many applications that extend far beyond it; this represents a fundamental development in thinking at the organizational level [81]. To provide mobility services, improve customer experience, optimize supply chain and production processes or improve the engineering and R&D of new products – you name it – AI is applicable to almost every area and can lead to untold opportunities to meet yet-to-be-defined consumer and partner needs. In the end, we are talking about a cultural change towards AI-driven organizations [82].

Being able to implement AI and having the capability to deploy at scale is a key-value differentiator in the auto industry. AI is a new fundamental way of treating and understanding data, opening many new possibilities in many realms, but the scale needed for it to succeed is significant when we talk about AI applied to vehicles in the road. Try to imagine, for example, the magnitude of the infrastructure required to learn from a growing vehicle fleet and then to update this fleet's software over time. This could entail thousands of vehicles and massive amounts of driven miles [83], all producing data continuously. Scale has an important impact on the tools and infrastructure to be deployed; these should put automakers at the forefront of innovation and competitiveness by allowing the relevant amounts of data to be properly managed, analyzed and understood. For this purpose, the auto industry needs teams with expertise in data science: teams that can implement new AI methods to best utilize the mountains of data continuously generated by their customers, the market and the internet. The industry needs specialized teams that can develop and implement tools to support a data scientist's work; it needs infrastructure ready that can scale up from a small, pre-development Proof of Concept (POC) to large volumes of data for mass production. Larger volumes of data not only demand smarter algorithms, but also more compute power, larger data storage, more powerful datacenters and innovative data management tools, as well as more sophisticated ways to measure performance, in order to quickly assess whether the development being done is going in the right direction.

This is no longer a case of human-defined software with hand-crafted features; instead we're looking at machine-defined software and automated data-driven feature selection. Again these are software-defined systems, as opposed to systems from previous generations where the central focus was on the hardware. Data is extremely valuable, and even more so if it is properly understood and labeled with attributes that give it further meaning and provide valuable insights into what that data really means. This implies the need for smart, well thought-out solutions to obtain such data, including for example relevant data selection, pre-tagging during data collection campaigns, and automatic and semi-automatic labeling tools.

Data collection is an essential aspect; limited amounts of data mean a kind of blindness and therefore reduced capability to improve forward. Data is key, and the process of acquiring it is very expensive and extremely important. There is extensive research and development focusing on obtaining data through simulation. To do the latter, one needs proper understanding and accurate modeling of the data being treated; AI and ML can provide benefits in both cases. Another key aspect is how to verify and test the validity of any ML/AI algorithm — especially when these algorithms need to be deployed in the field in safety-critical situations. Such verification and validation checks are usually carried out using immense amounts of data [83].

In the following pages, we will give some examples of how code can be derived from data in a software-defined approach; we will walk through how simulation can leverage valuable data; and we will cover how testing can be complemented with that synthetic data.

AUTOMATED RULES GENERATION

Nowadays it is not common for a programmer to start writing code from scratch. There are already many building blocks, created by other programmers, which can be reused and leveraged. Similarly to building a car, where engineers can take advantage of different components and parts that already exist and that come with detailed specifications and notes on how to make use of them in next-generation vehicle design, a software programmer can also use what we call libraries to make use of existing pieces of software by employing comprehensive application programing interfaces (APIs) and application notes.

A programmer sometimes repeats the same tasks while reusing the same pieces of code, a practice that can be furthered by scripting new pieces of software. In this sense, the programmer writes pieces of software that can make sense of already existing pieces of code and transform them in a process that we could imagine as code writing other code. This doesn't mean that programmers are not needed anymore, nor that code itself can actually create or generate any sort of software. There is no AI today that can write algorithms and code to solve problems the same way a human programmer does. Today's artificial intelligence can predict mistakes in code, can learn patterns and be trained to detect objects in pictures or to make recommendations, but it cannot think and reason like humans do.

Let's imagine a Traffic Sign Recognition system as an example to explain the difference between human-written algorithms with handcrafted features and AI-learned rules. In this context, the word "feature" refers to shape, color or some sort of pattern which characterizes the object we want to detect and classify. The algorithms behind a traffic sign recognition system will look for those features, compute the likelihood of the object being a traffic sign, and assign the class to which this traffic sign belongs.

We – with a human eye and human logic – might design a Traffic Sign Recognition system looking at shapes: for example, circular for speed limits or restrictions, triangular for warnings and rectangular for highway or city entry (in Europe). The first step to find traffic signs in an image might then consist of finding those regions of interest (ROIs) with a high probability of containing one of those shapes. Another intuitive cue we might use as humans is color: for instance red for speed limits, white for end of limitation, blue for recommendations and yellow for construction zones (this color assignment is an example; notice that colors have different meanings in different countries). We might also look for a diagonal stripe crossing the traffic sign to indicate end of limitation; or a narrow, horizontal white rectangle inside a red circular sign indicating "Do not enter". One further step might be to identify a number, for example within a speed limit sign. In this case, we might define certain patterns that help recognize numbers. All these features – shapes, colors, patterns – are chosen by our human intuition of what the most distinctive points of each traffic sign are, in order to help us create an algorithm to detect and classify them. Most of these features belong either to the so-called histogram of gradients (HOG), to Haar-like features (digital image features used in object recognition) or to any low-level feature set such as edges or corners. The main limitation of this kind of approach is that there is no guarantee that the recognition of these features will lead to an optimal descriptor for a specific traffic sign.

A human-handcrafted set of features might be constructed following a certain human-defined sequence of rules. This sequence can be imagined as a decision tree. First, we find regions of interest that contain a certain shape; then we branch e.g. circles into maximum speed limit, end of limitation and minimum speed limit based on color; next, we branch even further into the numbers within them. Such an approach seems logical from a human perspective but it might not be the best path. For example, to recognize a maximum speed limit of 30 km/h, the algorithms will go through circle, red, maximum speed limit and finally the classifiers for 30 only because we have defined it like this. However, it might have been a better choice to classify first the color and then jump directly to the number, without the need to go through shapes. Different algorithm developers might come up with different handcrafted features and ways of combining them, yet maybe none of them would find the most optimal solution.

AI instead will learn and weight the most relevant features, based on training data. We might imagine a neural network with several hidden layers in between. It is in those hidden layers that a convolutional neural network finds the most useful features for that specific classification task; in the case of traffic sign recognition, it will search for the best features to classify traffic signs. As opposed to decision-tree-like human rules that humans can also understand, automatically generated rules represent a model of the network incomprehensible from a human perspective – especially the network configuration and what it gives weight to. Automatically training feature recognition is very effective for solving the specific task the features have been learned for; it outperforms approaches where features are selected manually. However, the features thus learned and recognized are mainly good for that specific task – i.e. do not generalize to solving other problems. This approach also requires a large amount of training data, and this data needs to have the right quality for each specific task. The process known as transfer learning can sometimes help here by leveraging data in an attempt to reuse pre-trained models; it also tries to extend these models' capabilities for detecting and classifying new classes by bringing additional data specific to the new task.

Different data samples can lead to different results, which leads us to the ongoing debate on AI algorithms' capability to consistently produce the same output from unchanged input, especially in safety-critical applications. The developer has no control over those selected features; therefore validation through testing is key to ensure that AI algorithms will perform reliably and robustly. We will cover some validation approaches in the following section.

VIRTUAL TESTING PLATFORM

We have described how software is becoming the cornerstone in the automotive industry. We talked about sophisticated, complex AI-driven software whose algorithms generate rules extracted automatically from data. Increased software and algorithm complexity impose higher requirements for software-testing approaches, not only to benchmark a specific algorithm but also to test overall functionality and the interoperability between different systems in a vehicle. AI will play an important role in achieving such testing capabilities [84], allowing for automated test-case generation, faster time to market, and more accurate and efficient testing; for example, an "adversarial" approach for identifying problematic test scenarios has been proposed for testing ADAS/AD algorithms [85]. Here the term "adversarial" refers to the idea of having two competing neural networks (two "adversaries") that train each other. The first network aims to achieve the task at hand – for example, vulnerable road user detection – while the second network conducts a test by attempting to "fool" the first one, generating synthetic data samples that e.g. look like real pedestrians even though they are not.

Similarly to how we talked about handcrafted vs. AI-learned rules, we can look at manually defined tests and humanly automated testing vs. AI-written and – optimized testing. Handcrafted testing requires large amounts of time – especially if one wants to achieve proper coverage (i.e. addressing all possible conditions) – and this is an increasing trend, given that software is continuously evolving. Such constantly changing software needs fast and scalable testing platforms, which however requires engineering time and computer power – in other words, large investment. Reviewing one line at a time might take days or weeks, which is untenable in a world where new data is constantly produced and hence new software is written daily.

In the early stages of a project, one might hope to test the initial pieces of code as they arise – meaning one test would be set for each new code block. This could sound doable if one has the discipline to write tests at the same time that new software blocks are created, but it becomes non-sustainable over time; since new code interacts with existing code, the testing complexity grows and at some point is no longer scalable. The moral of the story is that it is necessary to have testing capabilities that can evolve alongside code development.

In fact, a good testing strategy is mission critical – and not just for testing at the functional level (commonly known as unit testing) but also including integration, system and acceptance testing [12]. When testing at functional level we want to ensure that a specific piece of software is doing exactly what it has been programmed to do on its own; for example, if a software block is computing sums of two integer values, the unit tests will check that it does the sum correctly but will also check for boundary conditions (inputs that are false because they exceed the system's parameters – for example, if the input is a non-integer) and faulty conditions (programming mistakes). Integration testing instead will test the behavior of all components which are connected to each other; for example, the interfaces between software blocks need to speak the same language for them to work together. At the system level, the software needs to be tested together with the car's other systems, with the aim of ensuring that the whole package can work in all expected use cases. System-level testing also includes assessing corner cases and stress situations that could cause memory or compute overload. Finally, acceptance testing in the automotive field is conducted by means of test drives in different geographical, weather and even extreme climatic conditions.

Testing is very important, not only for a new software block but also for the maintainability of existing software and its possible interactions. Software modules might interfere with each other, either as an expected interaction or due to some unintended behavior. The latter can be caused when different pieces of software share the same resources, since these resources have not been properly allocated to each specific process – for example, compute power or memory bandwidth. Unintended behavior can also be caused by buggy software or misuse of APIs or interfaces between different software modules, or may simply arise because some software or its interfaces have evolved or been deprecated (made essentially obsolete), as might happen when newly updated software coexists with old software. Although this is not an exhaustive list, it illustrates how errors – sometimes critical ones – can originate from diverse sources, making it rather complex to identify, isolate and ultimately fix them.

In order to reach insightful test coverage and sufficient regression testing, the automotive industry implements the Software in the Loop (SIL) and Hardware in the Loop (HIL) verification and validations methods. SIL is used to validate the intended functionality of specific software modules on big servers or developer computers. Pre-recorded real and synthetic data is run through the software module being tested and its outputs are evaluated. If these outputs match expectations, the software module has been proved fully functional. HIL, on the other hand, is used to test at system level; in this case, the target platform – the final computer to be installed in the car – is used to evaluate the software. Testing on top of the target HW (i.e. using the hardware that will be in the vehicle) makes it possible to identify potential issues with memory, resource allocation and real-time inconsistencies. To give some examples of the latter, software components that work fine on the developer's desktop might not meet runtime budget (i.e. could take too long) on the final embedded HW; or the speed in which sensor data comes may be higher than the speed the final HW can process it. This is an attempt to test on the bench (that is to say, not in the field) exactly as though inside a real vehicle – which is no small task, since it requires emulating all input data in a timely manner in order to produce the same environment interactions and system reactions that a real vehicle would. In the next section, we will go into more detail about simulation and AI, but one point for now: as one can imagine, both SIL and HIL testing require the proper infrastructure, which could mean large investment.

Connectivity offers another possibility for testing that involves what is sometimes called "shadowing" or "shadow mode", which runs directly inside the vehicle while it is on the road [86]. Shadow mode requires the computer in a vehicle to have enough capacity to run certain functions in the background while at the same time running its user-exposed functionality. Ideally, one could imagine a vehicle capable of running twice its full software stack: both the actual ADAS/AD software that is driving the vehicle and a new version of that software that is currently under test. In reality, though, compute power and resources might not suffice to run twice the amount of software concurrently. For this reason, it is necessary to choose strategies to optimize shadow-mode benefits, for example selecting relevant SW components to run and capture data on the vehicle while computing heavier post-processing in the cloud. Having the right heuristics in place can trigger onboard data logging (data recording) following certain events, for example after comparing the outputs from the software being tested with driver behavior and/or with current "foreground" software outputs. The scale of such an approach grows with the number of vehicles on the road enabled with this capability, which becomes a powerful asset to identify corner cases, improve development and testing, and reduce time to market of new and improved functionality.

Take for example an automatic emergency braking function, which the user is aware of and can benefit from. Now imagine that a new software version of that function has been developed and needs further testing. In a shadow-mode approach, the new function will be running in the background – without the user even knowing that it is there and without producing any interaction with the mechanical braking system. This new software under test will receive all the sensor inputs and run its corresponding analysis, producing a certain output – for example imminent breaking action – although that action will not be applied. The output of these background functions will be further analyzed, maybe through comparison with the actual version of AEB on the car or with the driver's current action. This comparison is useful both to generate certain test results for the new function and to identify situations that might not have been foreseen, which will be very valuable for further improvement and development of that function. Shadowing at scale will produce large amounts of data, requiring further data analytics that AI can certainly help with – not only when analyzing the data, but also when choosing what might be the best tests to obtain relevant information.

SYNTHETIC SCENARIO GENERATION

We have talked about shadowing, SIL and HIL; all of these require sufficient amounts of data to produce useful information. In this section, we will dive into why simulation, as an additional means of acquiring data for testing and development, is becoming a key element in the field of ADAS and AD.

In automotive, simulation has been widely used to model aerodynamics, material resistance, vehicle dynamics – and the list goes on. It has gained even more relevance in recent years with the explosive growth of ADAS and AD technology [87]. In this context, simulation becomes relevant for two main use cases: algorithm development, and verification and validation. Through simulation, developers can generate a vast array of scenarios, having full control of the environment and traffic situations, and even can randomize different daylight and weather conditions. This is similar in concept to a photorealistic racing video game: the scene is rendered by combining computer graphics, robotics dynamics (concerning forces applied) and kinematics models (concerning motion), trying to make things as realistic as possible by staying within the laws of physics. As an illustration of the interest in simulation for ADAS/AD, here are some open-source simulators: CARLA, AirSim and Deepdrive; the commercial solutions include ANSYS, dSPACE, PreScan, rFpro, Cognata, Metamoto and NVIDIA DRIVE Constellation simulation platform, just to name a few [88].

ADAS/AD systems must pass through extensive verification and validation in order to achieve FUSA (functional safety) and SOTIF (safety of the intended function) requirements (see AI for Safety section in Chapter 6 for more details). These requirements can only be fulfilled when the system being tested is proven to work over millions of kilometers. Acquiring such vast amounts of data is a complex and very expensive task, nowadays done via hours and hours on the road collecting data across different geographies, road types and weather conditions. As explained in the previous section, one possibility for generating data is to provide the vehicle fleet with shadow-mode capabilities. In this specific use case, shadow mode is employed to collect relevant data from a vehicle fleet that is already in circulation and equipped with some level of ADAS – of course including sensors and an ECU. One possibility is to trigger data logging when certain criteria are met, for example in use cases where developers lack data concerning specific geographies, weather conditions or driving scenarios. Shadow mode will log sensor data on those situations and put it in the onboard storage of the vehicle; once the vehicle reaches home or some connectivity point, this data is uploaded so that algorithm and function developers can make use of it.

However, even with data collected over hundreds of thousands of miles, there are certain circumstances and scenarios that will rarely be captured, for example, non-compliant behavior of other road users, or accidents. Yet such scenarios are necessary in order for smart ADAS/AD algorithms to be trained to handle such situations. Simulation is a possible solution to the problem of obtaining knowledge of these scenarios. Simulation may in effect save millions of dollars in driving cost, yet it is also not a trivial task, primarily because one must optimize the process of selecting relevant scenarios that cannot be easily found within real data. To do this, one must focus on generating those scenarios that impact most positively in the overall FUSA and SOTIF; as an example, [89] talks about how to select the right scenarios with the help of AI.

Likewise, simulation is a powerful tool for AI developers in training intelligent models and supervising the learning of their AD algorithms. In this context, AI algorithms can improve by using simulated data and simulation can benefit from AI. Simulation will benefit from AI through algorithms that allow better sensor modeling, photorealistic rendering [90] and material interactions closer to reality – in other words, the parametrizing of physics models during scenario generation stages. AI is also a valuable aid for simulation tools in accurately defining vehicle dynamic models (which are ways of describing a vehicle's motion – its roll, pitch and yaw – in addition to its progression in space) and when generating more representative traffic models [91].

We talked about SIL using recorded data to evaluate isolated SW components; such evaluation can also be done by replaying simulated data instead. Simulated data brings one important advantage over real data: it is by definition already labeled. In other words, since simulated data is artificially generated its characteristics are well known; hence its ground truth (what can be known about it through direct observation) is known and can be directly used to evaluate key performance indicators of that specific SW component. For example, our simulator may render a highway use case including various lanes and with other vehicles around what is called the ego vehicle (the vehicle from which the scene is being observed). In this case, the geometry (shape and relative volume), relative positions and speeds of all other agents are known and can be directly compared to the output given by the vehicle detector or lane detector algorithm.

By contrast, if we think about HIL, simulation brings one additional benefit by allowing closed-loop validation. Remember we talked about HIL in the previous section, referring to a means of testing in which we attempt to emulate the real system in the vehicle – which implies having the exact same computer, sensor inputs and actuator outputs. Closing the loop in this context means that one can test the system in real-time by feeding it certain synthetic inputs, and using that system's outputs to immediately define the next synthetic inputs. Imagine a racing video game where the player observes the scene and sends steering, acceleration and braking commands to the vehicle. If we replace the player with the ADAS/AV software stack we want to test, the computer receives rendered sensor inputs (camera, radar, lidar); the ADAS/AD algorithms perceive the scene and generate actuator control signals such as steer, brake or accelerate. These signals are then sent back to the simulator, which will compute and render the next sensor data. Hence the loop is closed, as would happen in a normal driving situation in which the driver continually perceives the situation, acts in response, sees the results of these actions, responds again etc.

Like simulation, synthetic scenario generation has been used extensively in the automotive industry, for example in the field of ADAS verification. This is typically done using special simulation software; a test engineer models the virtual 3D world including type of road (for instance highway or city), road signs, other road users etc., as well as the scenario (for example a lane change or parallel parking). The software simulates the raw sensor data based on the sensor models and locations in the vehicle as well as other necessary signals (car velocity, GPS coordinates etc.) and feeds them to the ECU being tested. This ECU receives this input data and processes it as if it came from a real vehicle operating under real driving conditions. As sophisticated as the simulation software already is, however, a lot of manual effort is required to set up, model and maintain these synthetic scenarios. Yet as we shall see shortly, AI has the potential to become a game changer in the realm of synthetic scenario generation as well.

Generative Adversarial Networks (GANs) undoubtedly represent one of the most innovative developments in recent AI. First proposed by Goodfellow et al. [92], GAN technology has catapulted AI's possibilities to a new level and catalyzed the development of a wide array of AI applications including super resolution (creating a high-resolution image from a low-resolution one), face synthesis (generating a realistic face image of a non-existing person), text to image (producing an image from a text description) and many others.

GAN is a framework made up of two independent deep neural networks, namely the generator network and the discriminator network. These two networks have opposing goals: while the generator network aims to produce realistic samples (image, speech etc.) in order to "fool" the discriminator network, the latter's goal is to distinguish whether those samples are fake or real. Both networks are trained to compete with each other until a kind of saturation point is reached. At this stage, the artificially generated samples become so realistic that the discriminator can no longer tell whether they are real or fake.

The GAN specialty of producing impressively realistic artificial data makes this a promising technology for synthetic scenario generation. Realistically generated scenarios are not only important for saving development costs but may be the only way to train and verify a car's behaviors in response to all possible situations. These might include rare circumstances or hypothetical cases which are not often encountered during real test drives and data collection sessions but which are necessary to prove that a vehicle would still operate correctly and safely if these unlikely events occurred.

Several studies investigating the performance of GAN-based frameworks in generating synthetic scenarios have shown promising results. Liu et al. [93] proposed the Unsupervised Image-to-Image Translation Networks (UNIT) framework capable of generating visually realistic, modified (day to night, snowy to dry road etc.) street scenes out of real driving video recordings. The Pedestrian-Synthesis-GAN (PS-GAN) framework proposed by Ouyang et al. [94] is able to embed sharp and photo-realistic synthetic pedestrian images in real scenes; these can be used to further improve the performance of the CNN-based object detectors that perceive them. Besides visual (i.e. image- or camera-based) synthesis, synthetic data generation from other sensors has also been investigated, for example using the data presented by radar signals [95] or lidar point clouds [96].

SUMMARY

The increasing relevance of AI in automotive R&D has been discussed in this chapter. AI brings a new mindset, which implies changes at the organizational level; now the organizational strategy needs to be consonant with the expertise of development teams as well as in line with the infrastructure surrounding the AI components. This is a fundamental change driven by data (specifically, the presence of massive amounts of data which of course require processing); it brings new opportunities but also poses new challenges concerning how to obtain and manage relevant data and to develop optimal algorithms. Also, once new algorithms are developed, how does one test them and then deploy them at scale?

First, we discussed the concept of automated rules generation, which engineers use to accelerate and optimize software development. In this context AI is used in development tools to predict errors in the code, to learn patterns, and to automate and speed up development, all of which turns into shorter time to market. AI is also used to select features for certain tasks, for example in image recognition, where less reliance on handcrafted features leads to higher recognition rates. Since AI algorithms are mostly data driven, one of the challenges is how to acquire reliable data that is relevant to the task at hand.

We then talked about the complexity of testing AI software, for example by making use of existing tools that at least partly employ AI, or by bringing new AI concepts to test everything from small software components to a full driving-software stack within a running fleet of vehicles. This is where we referred to the shadow-mode approach, which runs these SW components in the background for the purpose of testing and training them. Shadow mode also offers an effective way to obtain data for the further improvement of existing functions and the development of new ones. Once again here, data is the cornerstone. Through it one can not only identify weaknesses but also implement improvements, deploying them in the field via over-the-air updates. This method of diagnostics and improvements is what we call a software-defined approach. Having the right infrastructure in place is one of the general challenges in AI – affecting software testing, data acquisition and management and software deployment at scale.

To repeat, data is key. Simulation is useful to produce new data, which by definition comes paired with its ground truth since the features of synthetically generated scenarios are already known. Simulation is especially relevant for the study of corner cases such as accidents or non-compliant behaviors, since these are otherwise difficult to capture while driving. Simulation relies on modeling reality – creating models that imitate the exact characteristics of the sensors' input concerning the physical environment and the vehicles, traffic situations and so on that it contains. Achieving models that can recreate reality up to the level that is useful for development and testing of ADAS/AD algorithms is one of the main challenges within existing simulation solutions. Finally, we saw that GANs have the potential to become a game changer in the field of synthetic scenario generation, as they have shown impressive results in generating visually realistic artificial scenarios for use in automotive simulation.

5

AI FOR SERVICES

Putting the automobile industry and Artificial Intelligence (AI) together in the same sentence will quite likely bring to mind the term
Autonomous Driving (AD). It is a fact that auto manufacturers and
high-tech companies are racing to bring AI and self-driving technologies to the market; what's more, the application of AI in the
automotive industry extends to several domains, moving from
purely mechanical processes to more intelligently managed ones in
manufacturing, quality control and logistics. AI is also relevant to
various big-data-based services – encompassing cloud services,
procurement and finance, after-market services and digitization of
all processes within the vehicle's life cycle; and it extends to personalized marketing, fleet management, mobility and traffic in
smart cities and predictive diagnostics and maintenance.

Today's vehicles generate significant amounts of data because of onboard telematics monitoring (involving technology from telecommunications, electronics and software), driver assistance systems' sensors, driver behavior monitoring and other sources. At the end of the day here we're talking about big data which, combined with advanced data analytics, can provide rich insights to Original Equipment Manufacturers (OEMs), helping them not only to improve their internal processes but also to increase customer satisfaction. OEM-internal processes such as production, quality control, logistics and inventory greatly benefit from ML/AI algorithms. A good example on the production line is the use of a series of signals to train LSTM (Long-Short Term Memory, as you may recall) neural networks to help predict cracks during the manufacturing of metal sheets, allowing intervention before these cracks actually occur [97].

This data, coming directly from the vehicle itself, together with data from e-commerce channels or social networks, as well as other third parties such as repair shops, insurers and so on, can be used to learn customer preferences and needs. This is very valuable not only for the design and manufacturing of new models; it also helps aftersales teams fine-tune their services and operations to better capture customer needs, with a boost in customer loyalty and in sales as the ultimate goals. Directly related to this, responsiveness in case of a breakdown is much faster when the vehicle is connected with a digital supply chain that can anticipate what parts might be needed. We will have dedicated sections on predictive diagnostics on board the vehicle, running on-line analysis of vehicle data to assess the current vehicle status; and on predictive maintenance making use of various sources of data to help with continuous vehicle maintenance far beyond scheduled checks – a process which improves vehicle operation as well as workload distribution in service workshops.

Connectivity and digitization enable a wide spectrum of new innovative customer services which can be deployed through the infotainment system in the cockpit. Such services benefit from what is called the Internet of Things (IoT), in which many actors are connected to the cloud, including other vehicles, personal devices (e.g. cell phones), network-connected roadside units (such as traffic lights) as well as traffic-monitoring cameras and embedded road sensors [98]. All this information is processed either in the cloud, in the edge (this is understood in terms of distributed computing. One could think of intermediate computing stations between the cloud and the vehicle, with the result that data is not transmitted to or processed in the cloud; instead this happens closer to the location where it is needed, improving response time and bandwidth needs) or in the vehicle itself. The IoT is the enabler for services that assist the driver by providing real-time information or perhaps simply some new entertainment possibilities. The services on offer include real-time navigational direction with current traffic information; weather information and forecasts throughout the route; probability and prediction of free parking spaces near the destination; voice control and voice assistant; and supplementary road information provided by augmented reality. And there are already many more of these services now being developed.

AI will reshape the way we understand mobility inside urban areas, creating what we call smart cities, by reducing congestion and pollution and by providing faster, better, cleaner and cheaper means of transportation [99]. Mobility as a Service (or MaaS – namely the integration of digital platforms with various forms of transportation to make mobility an on-demand service – including trip planning and booking, car sharing and more) is being actively discussed, and without a doubt AI will be a key enabler here. Some existing AI algorithms within the context of smart cities and MaaS are Artificial Neural Networks (ANNs) and Genetic Algorithms (GAs), both of which can solve complex optimization problems within urban networks. Swarm intelligence methods have also been explored, taking inspiration from how ants or bees work together. These methods include the Ant Colony Optimizer (ACO) and Bee Colony Optimization (BCO) algorithms [100]. This technology will benefit transportation companies as well as car-sharing and ride-hailing platforms, all of which require smart fleet management services. Optimizing of travel time and distance, passenger waiting time, pick-up locations and vehicle capacity are all in sight here; in short, efficiency and quality of service will be improved.

Insurance companies certainly play a significant role among the many automotive service providers. Modern insurance models are evolving towards flexible and custom offerings, tailored to each driver and directly derived from their driving behavior. We will elaborate further on customer behavior and its correlation to insurance models later in this chapter.

PREDICTIVE DIAGNOSTICS

Modern vehicles are equipped with more Electronic Control Units (ECUs) and software per vehicle than ever before. Some of the functions these ECUs monitor are safety critical, giving self-diagnostics a very important role in ensuring that all systems are up and running as expected. Self-diagnostics in cars is not a new concept; sensors informing about possible engine failure, a low oil level or even low tire pressure were first incorporated into cars many years ago. However, self-diagnostics is becoming a more complex data analytics topic with the introduction of more sophisticated electronics, sensors and software – as well as the fact that vehicles are now connected rather than isolated entities. Self-diagnostics might help prevent and even fix some issues on the fly, through over-the-air updates, without the need for recall; they might even help extend the useful lifetime of a vehicle.

Onboard self-diagnostics benefits from the fact that it has real-time access to in-vehicle data streams and can sample as often as necessary, which can help quickly assess faulty situations at earlier stages. However, the ability to diagnose fast and reliably is directly dependent on the available data-processing power. Onboard storage and compute capabilities have traditionally been limited, which means a relatively small data-analytics capacity. The trend toward higher-performance ECUs on board the vehicle opens the door to more advanced data-driven approaches, which might be better suited for detecting and isolating future faults. If we focus on real-time, onboard-only data-stream analytics, we see methods that stress estimation of the Remaining Useful Lifetime (RUL), whereas other methods try to detect deviations from what could be considered normal operational behavior [101].

It is becoming possible to link these two approaches. The literature shows that Recurrent Neural Networks (RNNs) can be used to monitor the health of "time series" data (data perceived and plotted in its temporal order) coming from various sensors [97]. RNNs combined with autoencoders learn normal behavior from these temporal series in an unsupervised manner – i.e. without any previously structured definition of what they are looking for. The reconstruction error of new time series is then used to assess the degree of degradation of a machine, which is directly linked to its RUL. One can implement algorithms to detect deviations of temporal series by employing typical methods with names like sensor signal correlation matrices, linear regression, k-nearest neighbor clustering or fuzzy logics [101]. Detecting deviations might flag out that there is a trend towards a future fault, but it might not be so easy to isolate the root cause for such a deviation. In order to do the latter, historical data of similar deviations linked to known failures is needed – allowing the mapping of the currently detected deviation in relation to previously learned models, hence identifying the possible issue at hand.

In the context of ADAS and AD systems, knowing the health status of perception sensors is a must. Machine-learning techniques are used to identify and classify possible issues that impede a sensor from performing as intended. For example, soiling detection on cameras, also known as blockage or occlusion detection, can be done with Support Vector Machine (SVM) classifiers. Generative Adversarial Networks (GANs) can also be used, for example, what some authors have called SoilingNet [102]. Imagine a situation where dust, mud, ice, water drops or condensation partially or fully cover the lens. The results could be a limited field of view, reduced obstacle and object detection accuracy – or even complete unavailability of the latter feature, as if there was no sensor. Employing machine-learning techniques can not only inform the ADAS/AD algorithm about limited functionality but also help apply the appropriate countermeasure. For example, if the lens of a camera is found to be covered with ice or condensation, a heating mechanism might be activated, whereas if mud or dust is detected instead, a cleaning system could help deal with the problem.

Self-diagnosis of sensor availability and the application of corrective measures during operation could easily apply to other components, from memory subsystem integrity within the ECU through vehicle network and in-vehicle communications, and on to mechanical issues on the engine, tire pressure etc. The flow of the process in most cases starts with sensors continuously monitoring one or several parameters of a certain component; this sensor information is transmitted to a control unit which estimates the current status of that system. In some cases this estimate will be used to adjust some setting in real time, as we saw with sensor heating and cleaning; on some other occasions, this will be used to notify the driver to take some corrective action, for example, to check tire pressure. Bringing connectivity into the picture, all this data generated within the vehicle can be transmitted to data centers. There the data will be analyzed together with information from other vehicles and historical data. The results could be used to identify and fix software bugs and later correct them through OTA (over the air) updates. This might also help optimize planning of maintenance tasks as part of "predictive maintenance" – which we will look at in the following section.

PREDICTIVE MAINTENANCE

In the automotive industry, the need for maintenance is relevant to all systems, robots and equipment involved in the manufacturing process, as well as to many components of vehicles already on the road. Traditionally, maintenance scheduling has been based on periodicity – servicing machines and vehicles on a regular basis and replacing parts at predetermined intervals. This is known as preventive maintenance. The trend now, however, is an evolution toward what we call predictive maintenance. Early maintenance approaches were reactive, either to equipment failure or the passage of a specified interval of time. Moving beyond that, automakers have started using certain metrics such as runtime to fine-tune maintenance intervals. Adding sensors to track performance of certain parts leads to a higher level of understanding of equipment status, allowing what has been referred to as condition-based maintenance. As the next step, condition-based maintenance goes hand in hand with AI to predict future equipment failures, creating what is known as Predictive Maintenance (PdM).

Nowadays, most of these production machines and vehicles are equipped with various sensors that actively monitor and transmit vital information such as temperature, humidity, workload, fatigue, vibration levels etc. These sensors produce considerable amounts of valuable data and logs. The combination of real-time data with breakdown and maintenance history across a whole fleet or multiple production lines can be used to identify conditions and patterns that could lead to failure before it happens, hence to predict when maintenance might be necessary and to prevent significant damage. Analyzing and modeling from available data, one can estimate RUL, identify fault root cause and implement fault prediction strategies. Besides this, one can optimize maintenance scheduling, either extending the working life of some equipment or preventing major collateral damage to other equipment, both of which can result in risk minimization and a reduction in repair costs, waste, production downtimes and associated losses.

Predictive maintenance makes use of recent advancements in big data technologies such as sensors, connectivity and data analytics to create one of the building blocks of what has been called Industry 4.0 [103]. Machine-learning algorithms are used to analyze data and make necessary decisions. These may either be well-known ML algorithms like SVM, DT or k-Nearest Neighbor (k-NN); or else more recent but still familiar algorithms based on deep learning

(DL), for example, autoencoders, RNNs with LSTM, GANs or Deep Reinforcement Learning (DRL). The literature is rich in both [104] traditional ML- and newer DL-based PdM algorithms, all designed to diagnose faults, predict RUL, learn health indicators and make maintenance decisions. Traditional methods have the ability to diagnose problems when the amount of available data is limited, while DL-based methods will learn and produce a more accurate evaluation of components' degradation and failure identification, provided that enough history data is available. In both cases, though, selecting the right information might be challenging. In one example [105], the authors analyze several parameters to monitor engine health and predict faults, monitoring for example ignition, fuel, exhaust and cooling. Sensor data is transmitted from the On-Board Diagnostics port (OBD-II) to a smartphone via Bluetooth and sent to the cloud for further processing. Principal Component Analysis (PCA) is used to reduce feature dimensionality, and four classification algorithms are compared: SVM, Decision Tree (DT), Random Forest (RF) and k-Nearest Neighbor (k-NN), where SVM is proven to outperform the other methods.

Predictive maintenance is very promising, but there are still challenges to overcome. PdM requires changes at the organizational level in terms of skillset needs, infrastructure and technology, all of which of course translates into more cost [106]. Depending on the complexity and potential benefits of implementing PdM, some companies might not have the financial power to implement it or might decide that the return on investment is not sufficient. Another challenge for PdM is data acquisition, management and processing. Defining what is the most relevant data for specific equipment, finding the right sensors and implementing the right processes to ingest data for machine-learning algorithm training is not a trivial task. Even when everything is in place and we can predict future failures, it might not be feasible to prevent them – or the mitigation measures might not be so different from what would be needed without PdM. In addition, technicians need training to be able to work with new maintenance methods.

DRIVER BEHAVIOR ANALYSIS

We have seen how in vehicle telematics, data processed through the OBD-II can continuously monitor real-time vehicle status, for example from the Controller Area Network (CAN) bus and the on-board motion and location sensors, which log vehicle speed, acceleration, direction, mileage and so on. This information is useful to derive driver behavior. Nowadays smartphones can provide one more source of sensory data – which is not directly linked to the vehicle but still is valuable for gathering additional GPS signals as well as acceleration and compass (magnetometer) data. Many cars today include a "black box" recorder, also referred as an event data recorder, which captures data that helps to reconstruct the scenario prior to an accident. Infotainment and driver-monitoring systems bring additional information. All these sources of data produce many helpful indicators of a person's driving behavior such as average daily mileage, driving hours and speed; variations in speed and acceleration; steering changes; ratio of rush hour to off-peak driving time; and time spent driving at night or under certain weather conditions, to name a few [107].

The analysis of driver behavior is relevant, including because it might have significant impact on costs. It could help reduce fuel consumption and thus also CO_2 emissions; this is especially relevant for heavy vehicles. It could also help correct bad driving habits that tend to produce more wear and tear on various mechanical components, hence extending the lifetime of these components and lowering maintenance and repair costs. In addition, it can help reduce the risk of an accident, thus lowering insurance costs – which leads us to the point that driver behavior analysis is of special interest for insurance companies. After all, human error is a significant cause of accidents – more specifically, errors occurring due to bad driving behavior such as drunk or aggressive driving; hard acceleration, braking and cornering; speeding; or simply distracted driving [108].

Insurance policies are typically based on factual figures: driver age, experience and history, together with vehicle brand, category and engine power, just to list a few. With this approach, insurance risk and thus cost is recalculated on a yearly basis. Customer behavior analytics introduces the opportunity to make insurance-update intervals and pricing more flexible by employing what is called Usage-Based Insurance (UBI). There already exist several UBI models which consider how often, where, when and how the customer is driving. UBI has been proven to benefit both insurers and drivers, not only by offering discounts based on driving behavior data, but also due to driving behavior improvement and accident reduction [109]. There are three types of UBI: Pay-As-You-Drive (PAYD), where the insurance premium is based on amount of distance driven; Pay-How-You-Drive (PHYD), where discounts might be applied based on a driving behavior score at the end of each trip; and Manage-How-You-Drive (MHYD), where the insurance price is calculated considering real-time alerts and suggestions that are given to the driver to encourage safe driving habits. MHYD is the most evolved model, in which not only a person's driving pattern is analyzed, but also their levels of drowsiness, fatigue and distractions. The vehicle will need additional data from driver-monitoring technology and/or access to physiological data from a wearable device, combined with machine-learning algorithms, to determine current driver state (refer to Chapter 1, where we described state-of-the-art driver-monitoring systems).

The evolution of smartphones motivates several novel driver-behavior monitoring approaches, thanks to the increasing accuracy of their sensors; in some cases, the smartphone is also used to collect data from OBD-II via a Bluetooth adapter. Machine-learning algorithms commonly used in this context are those fitted to identify patterns in temporal series (such as Dynamic Time Warping or fuzzy logics), or those that can model probability within a certain class (be it the probability of a good driver, the likelihood of an accident, pass/fail in terms of specific behaviors etc.), such as logistic regression to predict probability and severity of accidents given a set of driver behavior feature statistics [107]. More complex ML algorithms are benchmarked by [110], to assess aggressive driving from braking, acceleration, turning and lane-change data. In this case, the authors state that RF outperformed SVM, Bayesian networks and an ANN. Another example [111] shows how SVM can help detect abnormal driving by using acceleration and orientation information from a smartphone to identify specific types of vehicle actions such as weaving, swerving, skidding, fast u-turning, under- or oversteering in turns, and sudden braking. In [112], a set of physiological and driving performance features are used to train an SVM classifier, which is then used to distinguish drunk driving from normal driving. Recent approaches make use of Deep Neural Network techniques such as RNNs [113]. Long/Short-Term Memory (LSTM) and Gated Recurrent Unit (GRU) are used to capture and correlate irregular driving behavior features from several CAN messages; more precisely, they analyze engine, fuel and transmission data.

The first challenge that driver behavior analysis faces is similar to most data analytics problems, in which the difficulty is the selection of the most insightful data for that specific task. Selecting the right data source is beneficial for obtaining better results but also for reducing data dimensionality; the latter is especially important if we consider that the amount of data to be processed is increasing every day. Another significant challenge is the fact that these methods might deal with personal driver data, which could raise data privacy concerns; hence even if intuitively some data source may seem most appropriate for driver analysis, it might be best to avoid using the data it offers.

SUMMARY

In this chapter, we have seen how AI for services within the automotive industry encompasses a wide range of possibilities, from real-time navigation to optimized maintenance scheduling to customized insurance models. AI can be applied to services that might be visible to the end customer, but also any service needed through the life cycle of a vehicle, from the beginning of the supply chain to aftersales. AI-based services bring significant benefits to car manufacturers, suppliers, service providers and customers. Car manufacturers analyzing big data can learn customer preferences; they can also improve component durability and reduce repair or maintenance time through predictive diagnostics. At the end of the day, this translates into increased customer satisfaction and loyalty.

We have also seen how predictive diagnostics can be used to assess component status on the fly and in real time, for example, to appraise the health of the engine or to communicate when sensors that are critical to the task of ADAS or AD are not working properly due to blockage. Such assessment is possible since we now have more sensors monitoring various components on a vehicle and producing continuous data streams; more powerful control units which can process larger amounts of data; and advanced data analytics methods which can make sense out of that data.

Predictive Maintenance (PdM) represents one further step, where connectivity allows diagnostics data from many vehicles to be transmitted to the cloud. PdM then learns from historical data in order to predict the RUL of the vehicle's components and to optimize maintenance cycles. We have seen how traditional machine-learning methods, for example, SVM, DT and RF, are typically used when available data is limited. On the other hand, deep learning algorithms are employed if the amount of data at hand is large enough, this produces more accurate results.

Driver behavior patterns can now be learned with the use of big data from vehicle sensors, driver-monitoring systems and even cell phones. Many accidents occur due to driver misbehavior (aggressive, drunk or distracted driving, speeding etc.). Insurance companies are looking into new models that give drivers discounts based on their driving behavior; UBI integrates real-time alerts and suggestions promoting safe driving in order to calculate insurance premiums.

6

THE FUTURE OF AI IN CARS

The future of the automotive industry will go hand in hand with Artificial Intelligence (AI). In fact, we can see that some cars already employ AI algorithms today in Advanced Driver Assistance Systems (ADAS) to process sensor data, understand the environment around the vehicle and act accordingly. In the third chapter, we noted that some AI-based voice-assistant systems – using AI to understand voice commands from a vehicle occupant and produce corresponding answers – are already on the market. These are just some examples illustrating the start of a new cooperative phase between car and high-tech companies.

The experience that automotive Original Equipment Manufacturers (OEMs) bring to the table is mostly what we could put into the categories of mechanics, dynamics and kinematics, material sciences and power expertise. For their part, tech companies bring software on top of new and advanced hardware concepts in computing, computer vision algorithms, machine learning (ML) and natural language processing, as well as know-how for handling data and building the right infrastructure to manage it at scale. The combination of these two worlds means significant opportunities to improve safety systems, from advanced driver assistance systems now to entirely autonomous vehicles later on. But not only safety stands to benefit from these developments; so does customer experience. Some examples here are improved infotainment and cockpit features including recommenders and voice assistants (as we saw in Chapter 3), and easier access to any mobility service.

The landscape of partnerships and collaborations between OEMs, robotaxi start-ups and high-tech software and hardware providers is continuously evolving; it's difficult to forecast which automotive OEMs might implement their own software solutions and which might partner with high-tech companies. Right now there are many OEMs (for example BMW, Daimler and Volvo), technology companies (e.g. Microsoft, NVIDIA and Waymo) and start-up players (for instance, Argo.AI, Aurora and TuSimple) employing AI in the automotive industry. Meanwhile, the complexity and cost of the autonomous driving endeavor might produce significant changes in the landscape through mergers, acquisitions and bankruptcies.

Some refer to these newly conceived vehicles as "computers on wheels" – in other words, as one more kind of connected player in the Internet of Things (IoT). Having seen how software is becoming the key added value for cars, in this book we have described this type of vehicle as the software-defined car. Moreover, this software evolves over time, bringing improvements and new features and extended vehicle value via over-the-air updates, similar to the way our smartphones receive updates and upgrades over time. The best contemporary example of Over the Air (OTA) capabilities for cars is given by the automaker Tesla [79], which supports OTA for infotainment, navigation and ADAS software. Other OEMs have also introduced OTA, but mostly limited to maps, navigation and infotainment.

OTA seems destined to become standard equipment within the auto industry; it is one of the key enablers facilitating continuous safety improvement toward the goal of zero casualties. Although it may seem ironic, this stream of developments means that motor vehicles are not completely independent entities anymore. The future of cars and AI involves several building blocks working together, extending from the vehicle itself to computing in the cloud. Large infrastructure outside the car plays an important role ranging from data management, new algorithm training, navigation, vehicle-usage and traffic-data analytics to communication (V2X), as well as many other aspects.

Developments in privacy and cybersecurity are among the hot-button issues within the field of connected, software-defined vehicles. Cybersecurity mechanisms will need to evolve to keep up with new computing models such as quantum computing. Not being limited to the zeros and ones of classic computers, the quantum computer is able to process large amounts of calculations simultaneously. In fact, quantum computers will have the capability to process the same amount of data several orders of magnitude faster, theoretically making all present cryptography mechanisms and cybersecurity systems obsolete. Initial development and infrastructure investment costs for this technology might be prohibitively high for some car manufacturers, presenting another challenge to the extensive adoption of AI in the automotive industry.

In addition, AI-based self-driving cars add one more level of complexity when it comes to verification and validation. If one considers an end-to-end approach (one without human programming intervention) where the vehicle learns to drive by correlating sensor inputs with driver behavior, assessing the driving environment – and ensuring a suitable response to it – are no trivial tasks.

In the following sections, we will describe and compare two distinct machine-learning approaches which are already employed in the automotive industry: semantic abstraction learning and end-to-end learning. We will then go into some details that show how functional safety in cars can benefit from AI, as well as how AI algorithms help improve the intended functionality itself. Finally, we will go through some cybersecurity challenges for connected vehicles and will describe some opportunities AI brings in terms of security.

A TALE OF TWO PARADIGMS

ML and AI algorithms are both data driven, meaning that these algorithms learn certain models and patterns from data that is relevant to the task at hand. For example, driving-video sequences with all vehicles properly labeled are used to learn to recognize surrounding vehicles.

We have seen in earlier chapters how AI is used in the autonomous driving domain. In that context, we talked about using different modules to perform different perception tasks: one module for vehicle recognition, another to detect lane markings, yet another one for recognizing pedestrians and cyclists, and even segmentation DNNs to classify pixels into road, sidewalk or sky areas and so on. Each of these independent modules is designed and trained to perform a specific perception task. Then all these modules are merged to create a proper understanding of the surroundings and to further derive whether there are obstacles on the path, to predict if there are any possible crash trajectories and to generally plan where to go next. If we want to imagine the architecture of this sort of autonomous-driving software stack, we might visualize many specialized modules that have been taught and optimized for specific, individual tasks, and that work together to achieve a more complex, common goal. Yet in reality, given that compute power is bounded, some of these modules might be combined at earlier stages – for example, to identify regions of interest in order to reduce compute needs for those more specialized modules. However, these are still modular learning approaches, also referred as semantic abstraction learning [114].

By contrast, end-to-end learning within the autonomous driving context consists of deriving AD control commands by correlating sensory inputs with driver commands. In other words, the end-to-end approach learns to drive by imitating what human drivers do when faced with each given scenario. End-to-end learning makes it unnecessary to divide the task into human-defined sub-modules for specific tasks such as vehicle detection, lane marking detection and so on; therefore it is also not necessary to supervise each individual module. Instead, we talk about a single AI module. An implementation example of end-to-end learning is given in [115], the so-called PilotNet research project, in which the authors use a Convolutional Neural Network (CNN) in order to learn the entire processing pipeline for steering a vehicle by pairing the output from forward-looking cameras with (human) driver steering commands [116]. This approach demonstrates the ability to avoid obstacles and follow driving paths; but depending on whom the system is learning from, it may not learn to obey all traffic laws. In [117], conditional learning is added as a means to help the vehicle navigate towards intended intersections. The PilotNet approach has also inspired other methods, for example, a multi-modal sensor combination (lidar and camera) presented by [118], a lighter CNN model (which is computationally less expensive) presented by [119] and a surround-view camera end-to-end system presented by [120]. As we also have seen in the previous chapters of this book, the end-to-end learning approach is gaining increasing popularity in other AI applications as well, such as automatic speech recognition and driver's gaze estimation.

End-to-end learning is conceptually and mathematically beautiful, and in theory we can achieve our engineering goal by training a system in a unified manner, without worrying about what is happening in the system's guts. Conceptually, learning purely from data without the need for human experts designing sub-modules seems like a consistent approach [121]. And end-to-end learning has been proven to work in several domains. But within the automotive domain, where functional safety is a major factor, it still suffers from several limitations. The amount of data needed to achieve sufficient reliability in many scenarios and use cases is significant, which might make training and testing independent vision algorithms more efficient [122]. In fact, the vehicle would need to learn from many expectable use cases and scenarios – not only impacting the amount of data needed but also potentially requiring large neural networks, which for their part would of course need more compute power. In addition, a local optimum could be produced: overly limited data leading to an overly limited conclusion and finally, a complete learning failure. In the end, basically, the fact that we do not have full control of what is happening inside the learning process poses concerns on how to ensure consistency (getting the same output from unchanged input) and to verify and validate the algorithm.

As of today, end-to-end learning has been proven to work in certain limited scenarios within the automotive industry, but modular approaches still dominate in major ongoing developments and implementations. In the future, we might see hybrid approaches as a way to increase overall functional safety.

AI & CAR SAFETY

Although the automotive industry has evolved considerably since its beginning in terms of safety, the zero-fatality goal is still far off. Better overall performance brings higher vehicle speeds and new safety challenges, shortens the time window for reaction and means more severe impact when accidents do occur. The industry enforces implementation of safety policies all across the value chain to increase safety inside the vehicle, for passengers – and outside it, e.g. for vulnerable road users (sometimes abbreviated as VRUs). Modern vehicles are designed with a certain backup redundancy in critical systems, to ensure safe resolution of dangerous situations that can arise when failures take place. This is known as fail-safe operation; it brings the vehicle into a safe state, for example to a safe stop in its present lane. An alternative, known as the fail-operational approach, allows the vehicle to continue operating with reduced functionality and maneuverability until it can safely stop, for example on a highway shoulder.

Safety systems in the automotive industry can be divided into two categories: active and passive. Active safety systems are made to prevent accidents by providing assistance in steering and controlling the vehicle, while passive safety systems are targeted to mitigate the damage caused by an unavoidable accident. Many existing ADAS are active systems, for example, Automatic Emergency Breaking (AEB), Electronic Stability Control (ESC), Anti-Lock Braking Systems (ABS), Adaptive Cruise Control (ACC), Blind Spot Detection (BSD), to name a few. Some well-known examples of passive safety systems are the seat belt and the airbag.

Active and passive safety systems have gradually improved – in part thanks to government car-safety evaluation programs such as the New Car Assessment Program (NCAP) [123], which provides vehicle ratings based on how well a vehicle performs in adult- and child-occupant protection, pedestrian and cyclist protection and safety assistance. Although NCAP only gives non-binding assessments, it can be seen as an incentive for OEM investment in best safety practices. Outside the realm of governmentally sponsored programs, ISO defines two important standards concerning unsafe system behaviors in the automotive context: ISO 26262 Road vehicles – Functional Safety (FUSA) [3], and ISO 21448 Road vehicles – Safety Of The Intended Functionality (SOTIF) [124]. FUSA addresses unsafe behaviors when they are caused by system failure, while SOTIF deals with those caused by the intended functionality when the system does not fail.

ISO 26262 defines Automotive Safety Integrity Levels (ASIL), which help define safety requirements by analyzing risk of potential hazard severity, exposure and controllability. This stage of identifying possible malfunctions and assessing associated risk is known as Hazard Analysis and Risk Assessment (HARA). For example, an AEB system can generate a control action to engage the brake; one possible associated hazard would be that the brake is activated when it should not be. This control action will also have associated safety goals; for example, since the brake controller must only engage the brake when needed, there will be safety requirements for the sensors, electronics and mechanics so that each of the system components is conformant to FUSA [125]. FUSA identifies four ASIL levels, A (lowest level of integrity) to D (highest integrity), as well as QM (Quality Management) for hazards not linked with any safety requirements. Some good examples of ASIL-D-rated applications within a vehicle are AEB, ABS, ESC and airbag deployment.

Since ASIL-D-rated systems must ensure that there is no complete loss of functionality, the requirements for these systems' components (hardware and software) are highly demanding. Ideally, each element should be rated ASIL-D, but this would come at high development cost. ISO 26262 defines what is called ASIL-decomposition as a means to achieve the desired ASIL rating by employing redundancy and sufficient independence – hence diversity – for each of the elements contributing to a safety goal. In the AEB example requiring ASIL-D, the perception part could be achieved with multiple ASIL-B sensor modalities (camera, radar, lidar) checking for obstacles ahead. Having multiple sensor modalities brings redundancy – meaning that in the case of one sensor's failure, others are still covering the area being looked at; it also brings diversity, since each sensor modality performs best under different conditions. The same methodology could be applied to the software layer that detects the obstacles ahead, for example by running traditional computer vision (if you recall from Chapter 2, this is the science of trying to make computers visually perceive as humans do) as one path while running AI-based algorithms on a second path; these would be both redundant and diverse to each other and would complement each other to reach the same safety goal.

The standard known as SOTIF [124] complements FUSA and raises the bar for safety. All components in the system, be they physical (like sensors) or mathematical (particularly algorithms), must perform the task as designed under all conditions. SOTIF puts the focus on function completeness. If we take for example object-recognition algorithms, where the intended function is to recognize e.g. vehicles, lane markings, pedestrians and cyclists, the level of completeness for this intended functionality improves with the use of ML algorithms such as deep learning [126]. However, machine learning suffers from certain limitations and challenges within the safety context when it comes to safety certification. ML algorithms might have non-deterministic behavior, could be difficult to assess (especially given the limited visibility of what is happening within hidden deep-learning layers), and might produce instable results given the strong dependence on the particular data being used during learning stages [127].

Verification and Validation (VnV) also becomes more difficult with the use of ML algorithms. Typically, VnV processes make sure that software is free from bugs and that it follows coding guidelines and standards; however, VnV approval does not necessarily ensure expected behavior. Corner cases and rare scenarios might still produce unsafe behaviors. Deep-learning algorithms have pushed the limits of traditional computer vision algorithms since 2012 [44]. And although deep learning (DL) cannot replace computer vision (CV) for all tasks, it seems clear that DL provides substantially better performance than CV in terms of image processing, pattern recognition and scene understanding – which means an improvement in performance of the intended functionality under various conditions and hence an increase in SOTIF.

AI & CAR SECURITY

Cyber-security is becoming increasingly relevant in the automobile industry. Vehicles are gradually getting more connected to different external systems and actors, which increases the likelihood of cyberattack through the various vehicle communication systems (known collectively as V2X or Vehicle-to-Everything) [128]. These include Vehicle-to-Device (V2D) communication, for example, the infotainment system allowing the cell phone to play music or accept phone calls through the car's Bluetooth connectivity, or simply the remote keyless entry system. V2X also includes Vehicle-to-Infrastructure (V2I), used for example for vehicles to communicate with traffic lights, and Vehicle-to-Vehicle (V2V), in which vehicles exchange real-time positioning and speed information, for example in platooning use-cases (in which several vehicles drive very closely together as a group, by virtue of shared system information). This technology also enables a more general exchange of information with other nearby road users, e.g. about upcoming hazards. There's also Vehicle-to-Network (V2N), where vehicles exchange various types of data that is then stored and processed in a cloud-based infrastructure. There are even examples on how to communicate between vehicles and pedestrians (V2P) by using mobile apps to transmit certain signals in order to help prevent accidents involving pedestrians [129].

All these communication examples are possible access points for cyberattacks due to the vulnerability caused by connection to external counterparts (for example smartphones, the cloud or infrastructure), but the vehicle can also suffer cyberattacks that are channeled through internal systems. In fact any system that is hardwired to the driving ECU, including Ethernet, Controller Area Network (CAN), LIN, FelxRay or USB, could be a potential conduit for an attack. Take as an example any safety-critical vehicle control message, such as steering wheel angle or level of braking intensity, which needs to be authenticated (confirming that the message really comes from the purported sender) and verified (assuring that the message is not tampered with during transmission). Network messages can be tampered with in different ways. A malicious device can compromise the message, or spoofed information can be captured by a sensor – for example, fake traffic signs or road objects.

Taking a step outside the vehicle itself, one can focus on the entire digital ecosystem around it, where different software vulnerabilities are observable. These weak points could exist within OEM back-end services, where malware could be installed and expose vehicle data or even gain access to vehicle door control; third-party services such as EV (Electric Vehicle) home chargers could be accessed via home Wi-Fi, or for example through car-sharing apps; production and maintenance infrastructure can be infected with malware, causing significant disruptions in production lines and maybe even the stop of production across several plants [130]. Such vulnerabilities all across the ecosystem require a cybersecurity strategy for the entire lifecycle of a vehicle. For that purpose, regulations and standards are being defined. These efforts include the World Forum for Harmonization of Vehicle Regulations (WP.29) under the UN Economic Commission for Europe (UNECE) and ISO / SAE 21434 "Road vehicles – Cybersecurity engineering".

Both the UNECE and ISO upcoming standards put emphasis on "security by design" to cope with the increasing vulnerabilities caused directly or indirectly by networking on vehicles, by encouraging OEMs to have security in mind right from the early development phases. The goal is to have all hardware and software within the vehicle and its ecosystem designed, implemented and tested in regard to cybersecurity issues, and to ensure that the right mitigation measures are in place. Similar to the way that the ISO standards Software Process Improvement and Capability Determination (SPICE) and ISO26262 define both the "V model" (as the reference *process* model for the different phases of product development) and functional safety hazard analysis (HARA), in the context of risk management there is a "V" version for security. This method enforces continuous management of both software patches and overall vulnerability, with monitoring and response processes to face any attack. It also dictates a Threat Analysis and Risk Assessment (TARA) which is applied right from the beginning of any development lifecycle [131].

Various cyber threats can occur in the automotive domain [132], [133]: masquerading, where the attacker disguises pretending to be an authorized user to gain access to the system; eavesdropping, also known as a sniffing or snooping, where the attacker accesses data while it is being sent or received by the user through a non-secure network; spoofing, where the attacker impersonates legitimate devices or users in order to spread malware, steal data or even take control; and Denial of Service (DoS), where the system's bandwidth and resources are flooded, preventing it from performing its expected functionality and potentially even bringing on a system crash. Note that this is a sample list; there are many more existing cyber threats, and new ones appear fairly regularly. Any of them can lead to severe consequences directly affecting the driver and other road users due to system failures such as loss of brake or steering control. The so-called "Jeep attack" showed cybersecurity vulnerabilities resulting from an infotainment system's connectivity. The Jeep's externally connected infotainment ECU was also connected to the vehicle's internal network, and a security breach was exploited to take control of the car remotely. Other very negative consequences could include takeover of the Global Positioning System or personal data theft.

From an OEM perspective, attacks like this could easily damage the brand and translate into commercial losses. Their prevention starts with hardwired solutions, for instance, a Hardware Security Module (HSM). HSM is a form of dedicated security hardware that offers accelerated cryptographic calculations and better encryption processes through the use of true random number generators and secure key storage, among other things. However, hardware alone does not solve all security threats; additional supporting software components must be employed, potentially using AI and deep learning to enable authentication, digital signatures, Public Key Infrastructure (PKI) and secure boot and communication protocols [134].

In the context of cybersecurity, AI can be seen as a two-edged sword; it can certainly help protect against cyberattacks by automating repetitive defensive tasks, but at the same time it can be used by hackers to make far more complex and advanced attacks via deep learning, reinforcement learning, support vector machines and other ML algorithms [135]. We can separate protective cybersecurity approaches using ML [128] into the categories of supervised, unsupervised and reinforcement learning. Supervised machine-learning models can handle specific tasks after being trained with human-labeled data, but these are not generalizable – i.e. not applicable to other tasks. Some examples are Long/Short-Term Memory (LSTM) [136], Recurrent Neural Networks and Generative Adversarial Networks [137]. Unsupervised learning models are based on analyzing different non-labeled data streams in the vehicle and looking for anomalies and abnormal behaviors, for example, deep belief networks like deep autoencoders with hidden layers [138]. Reinforcement learning methods are very promising in terms of generalizing and providing a large variety of cybersecurity solutions, although this is the least proven approach. Various Deep Reinforcement Learning methods exist [139]; for example, the combination of LSTM, fully connected layers and regression, is used to defend against injection attacks – in which a hacker tries to inject faulty data into AV sensor readings.

The automotive industry will need new talent and skills to develop and implement secure software similarly to what's already being done in other industries such as aerospace, tech and critical infrastructure. Learning from these examples, together with following automotive standards, will set the basis for best practices for hardware, software and the full development lifecycle. There will always be significant challenges to keeping up with the latest security threats; this is especially relevant to products that have long life cycles.

SUMMARY

The future of the automotive industry will unfold together with software, be it AI-based software inside the vehicle to partially or fully drive cars or to assist drivers via natural language processing, or outside the vehicle to manage large-scale mobility services and logistics – or simply to analyze various sources of data that can help improve passenger safety and comfort – and thus also ultimately customer satisfaction. Software-defined vehicles will become standard, as will the infrastructure around them. A software-defined vehicle will require connectivity in order to receive updates over the air but also in order to send sensor and usage data to the cloud; hence, connectivity is key to further improving and developing features to increase customer satisfaction. Whether new features are learned in a modular way or an end-to-end manner, data is a very valuable asset. Data is necessary to understand where AI algorithms might need more development time, for example, given corner cases or simply to increase reliability in already significantly robust use cases.

AI algorithms are data driven and rely on abundant amounts of data to produce statistically relevant results; the latter is a must in order to achieve sufficient functional safety and to ensure that the intended functionality of each software component is achieved. In this context, we have discussed how recent DL algorithms outperform previous methods, for example in terms of image recognition, increasing safety. In addition, AI algorithms bring a diverse method to existing techniques, helping in what is called ASIL decomposition and allowing a system to reach higher ASIL ratings. It still remains challenging to verify and validate AI-based algorithms, especially if we talk about end-to-end solutions, but it is only a matter of time until we reach consensus on the right methodology for doing this.

A software-defined car is perhaps the most sophisticated object in the IoT world, always connected, which poses a major challenge in terms of anonymity and cybersecurity. We have seen how various cyberattacks could threaten the well-being of vehicle occupants as well as other road users. It is imperative to implement cybersecurity measures to protect against damage to software – and the rest of the system – that could come through over-the-air updates or other communication with the external world. A hacker could steal personal information or in a worst-case scenario even take control of the vehicle; therefore hardware and software cybersecurity (both in-vehicle and in the surrounding infrastructure) must be ensured throughout the whole product lifecycle.

Finally, we saw that AI in the cybersecurity context is a two-edged sword. AI is being used to detect and defend against cyber-attacks. On the other hand, it is also being used to discover new vulnerabilities and create more sophisticated attacks.

Despite all the challenges and unanswered questions, AI is undoubtedly playing an increasingly significant role in automobile technology. Given these developments, a better understanding of AI's potential can only help orient us better for the road ahead.

FURTHER READING

W. Burgard, S. Thrun, and D. Fox, *Probabilistic Robotics*. MIT Press, 2005.

J.-A. Fernández-Madrigal, *Simultaneous Localization and Mapping for Mobile Robots: Introduction and Methods: Introduction and Methods*. IGI Global, 2012.

D. Jannach, M. Zanker, A. Felfernig, and G. Friedrich, *Recommender Systems: An Introduction*. Cambridge: Cambridge University Press, 2010.

K. P. Murphy, *Machine Learning: A Probabilistic Perspective*. MIT Press, 2012.

S. Shalev-Shwartz and S. Ben-David, *Understanding Machine Learning: From Theory to Algorithms*. Cambridge University Press, 2014.

H. Sjafrie, *Introduction to Self-Driving Vehicle Technology*. CRC Press, 2019.

REFERENCES

[1] Edureka, "Artificial intelligence algorithms: All you need to know," 2020. https://www.edureka.co/blog/artificial-intelligence-algorithms/ (accessed Oct. 01, 2020).

[2] E. Sax, R. Reussner, H. Guissouma, and H. Klare, "A Survey on the state and future of automotive software release and configuration management," 2017. doi: 10.5445/IR/1 000075673.

[3] S. Singh, "Critical reasons for crashes investigated in the national motor vehicle crash causation survey," vol. 2018, Washing ton, DC: National Highway Traffic Safety Administration, no. April 2018, pp. 1–3.

[4] SAE International, "Taxonomy and definitions for terms related to driving automation systems for on-road motor vehicles," SAE Int., 2016.

[5] W. Hulshof, I. Knight, A. Edwards, M. Avery, and C. Grover, "Autonomous emergency braking test results," Proc. 23rd Int. Tech. Conf. Enhanc. Saf. Veh., pp. 1–13, 2013, [Online]. Available: http://www-nrd.nhtsa.dot.gov/Pdf/ESV/esv23/23ESV-0001 68.pdf.

[6] H. Sjafrie, *Introduction to Self-Driving Vehicle Technology*. Boca Raton, FL: CRC Press, 2019.

[7] A. Krizhevsky, I. Sutskever, and G. E. Hinton, "ImageNet classification with deep convolutional neural networks," *Adv. Neural Inf. Process. Syst.*, vol. 2, pp. 1097–1105, 2012.

[8] Z.-Q. Z. Wu, P. Zheng, and S. Xu, "Object detection with deep learning: A review," pp. 1–21, 2018.

[9] J. S. D. R. G. A. F. Redmon, "(YOLO) You Only Look Once," *Cvpr*, 2016, doi: 10.1109/CVPR.2016.91.

[10] W. Liu et al., "SSD: Single shot multibox detector," *Lect. Notes Comput. Sci. (including Subser. Lect. Notes Artif. Intell. Lect. Notes Bioinformatics)*, vol. 9905 LNCS, pp. 21–37, 2016, doi: 10.1007/978-3-319-46448-0_2.

[11] E. Yurtsever, J. Lambert, A. Carballo, and K. Takeda, "A survey of autonomous driving: Common practices and emerging technologies," 2019, [Online]. Available: http://arxiv.org/abs/1906.05113.

[12] L. C. Chen, G. Papandreou, I. Kokkinos, K. Murphy, and A. L. Yuille, "DeepLab: Semantic image segmentation with deep convolutional nets, atrous convolution, and fully connected CRFs," *IEEE Trans. Pattern Anal. Mach. Intell.*, vol. 40, no. 4, pp. 834–848, 2018, doi: 10.1109/TPAMI.2017.2699184.

[13] A. Paszke, A. Chaurasia, S. Kim, and E. Culurciello, "ENet: A deep neural network architecture for real-time semantic segmentation," no. June 2016, [Online]. Available: http://arxiv.org/abs/1606.02147.

[14] M. Siam, S. Elkerdawy, M. Jagersand, and S. Yogamani, "Deep semantic segmentation for automated driving: Taxonomy, roadmap and challenges," in *IEEE 20th Int. Conf. Intell. Transport. Syst.*, Yokohama, Japan, 2017, pp. 1–8. doi: 10.1109/ITSC.2017.8317714.

[15] A. Pfeuffer, K. Schulz, and K. Dietmayer, "Semantic segmentation of video sequences with convolutional LSTMs," *IEEE Intell. Veh. Symp. (IV)*, Paris, France, 2019, pp. 1441–1447, doi: 10.1109/IVS.2019.8813852.

[16] Y. Lyu, L. Bai, X. Huang, "Road segmentation using CNN and distributed LSTM," *IEEE International Symposium on Circuits*

and Systems (ISCAS), Sapporo, Japan, 2019, pp. 1–5, doi: 10.1109/ISCAS.2019.8702174.

[17] J. Wu, J. Jiao, Q. Yang, Z. J. Zha, and X. Chen, "Ground-aware point cloud semantic segmentation for autonomous driving," *MM 2019 - Proc. 27th ACM Int. Conf. Multimed.*, pp. 971–979, 2019, doi: 10.1145/3343031.3351076.

[18] J. G. López, A. Agudo, and F. Moreno-Noguer, "3D vehicle detection on an FPGA from LIDAR point clouds," *ACM Int. Conf. Proceeding Ser.*, pp. 21–26, 2019, doi: 10.1145/33 69973.3369984.

[19] F. Zhao, J. Wang, and M. Liu, "CDSVR: An effective curb detection method for self-driving by 3D lidar," *ACM Int. Conf. Proceeding Ser.*, pp. 38–42, 2019, doi: 10.1145/3325 693.3325695.

[20] B. Paden, M. Čáp, S. Z. Yong, D. Yershov, and E. Frazzoli, "A survey of motion planning and control techniques for self-driving urban vehicles," *IEEE Trans. Intell. Veh.*, vol. 1, no. 1, pp. 33–55, 2016, doi: 10.1109/TIV.2016.2578706.

[21] D. Connell and H. Manh La, "Extended rapidly exploring random tree–based dynamic path planning and replanning for mobile robots," *Int. J. Adv. Robot. Syst.*, vol. 15, no. 3, pp. 1–15, 2018, doi: 10.1177/1729881418773874.

[22] C. Xi, T. Shi, Y. Li, and J. Wang, "An efficient motion planning strategy for automated lane change based on mixed-integer optimization and neural networks," *Comput. Res. Repos.*, 2019, [Online]. Available: http://arxiv.org/abs/1904.08784.

[23] Z. Bai, B. Cai, W. Shangguan, and L. Chai, "Deep learning based motion planning for autonomous vehicle using spatiotemporal LSTM network," *Proc. 2018 Chinese Autom. Congr. CAC 2018*, pp. 1610–1614, 2019, doi: 10.1109/CAC.201 8.8623233.

[24] X. Jin, G. Yin, and N. Chen, "Advanced estimation techniques for vehicle system dynamic state: A survey," *Sensors (Switzerland)*, vol. 19, no. 19, pp. 1–26, 2019, doi: 10.3390/s19194289.

[25] S. Grigorescu, B. Trasnea, T. Cocias, and G. Macesanu, "A survey of deep learning techniques for autonomous

driving," *J. F. Robot.*, vol. 37, no. 3, pp. 362–386, 2020, doi: 10.1002/rob.21918.

[26] International Organization for Standardization, "Road vehicles — Functional safety (Standard No. 26262:2018)," 2018. https://www.iso.org/standard/68383.html (accessed Aug. 09, 2020).

[27] A. Greenberg, "Hackers remotely kill a Jeep on the highway—with me in it," *Wired*, vol. 7, p. 21, 2015.

[28] A. Fischer, "Bosch and Daimler obtain approval for driverless parking without human supervision," Press Release, Jul. 23, 2019.

[29] C. Unger, E. Wahl, and S. Ilic, "Parking assistance using dense motion-stereo," *Mach. Vis. Appl.*, vol. 25, no. 3, pp. 561–581, Apr. 2014, doi: 10.1007/s00138-011-0385-1.

[30] F. Abad, R. Bendahan, S. Wybo, S. Bougnoux, C. Vestri, and T. Kakinami, "Parking space detection," in *14th World Congress on Intelligent Transport Systems, ITS 2007*, 2007.

[31] F. Ghallabi, F. Nashashibi, G. El-Haj-Shhade, and M.-A. Mittet, "LIDAR-based lane marking detection for vehicle positioning in an HD map," *2018 21st International Conference on Intelligent Transportation Systems (ITSC)*, Nov. 2018, pp. 2209–2214, doi: 10.1109/ITSC.2018.8569951.

[32] J. Petereit, T. Emter, C. W. Frey, T. Kopfstedt, and A. Beutel, "Application of hybrid A* to an autonomous mobile robot for path planning in unstructured outdoor environments," in *Robot. 2012; 7th Ger. Conf. Robot.*, 2012.

[33] O. Amidi and C. Thorpe, "Integrated mobile robot control," Fibers' 91, Boston, MA, 1991.

[34] S. Thrun et al., "Stanley: The robot that won the DARPA Grand Challenge," *Springer Tracts Adv. Robot.*, 2007, doi:10.1007/978-3-540-73429-1_1.

[35] J.-A. Fernández-Madrigal, *Simultaneous Localization and Mapping for Mobile Robots: Introduction and Methods: Introduction and Methods*. IGI Global, 2012.

[36] R. Smith, M. Self, and P. Cheeseman, "Estimating uncertain spatial relationships in robotics* *The research reported in this paper was supported by the National Science

Foundation under Grant ECS-8200615, the Air Force Office of Scientific Research under contract F49620-84-K-0007, and by gene," in *Machine Intelligence and Pattern Recognition*, 1988, pp. 435–461.

[37] M. Montemerlo, S. Thrun, D. Koller, and B. Wegbreit, "FastSLAM: A factored solution to the simultaneous localization and mapping problem," in *Proc. Nat. Conf. Artif. Intell.*, 2002.

[38] G. Grisetti, R. Kummerle, C. Stachniss, and W. Burgard, "A tutorial on graph-based SLAM," *IEEE Intell. Transp. Syst. Mag.*, 2010, doi: 10.1109/MITS.2010.939925.

[39] N. Fairfield and C. Urmson, "Traffic light mapping and detection," *Proc. IEEE Int. Conf. Robot. Autom.*, 2011, doi: 10.11 09/ICRA.2011.5980164.

[40] L. Zhou and Z. Deng, "LIDAR and vision-based real-time traffic sign detection and recognition algorithm for intelligent vehicle," in *2014 17th IEEE Int. Conf. Intell. Transport. Syst.*, ITSC 2014, 2014, doi: 10.1109/ITSC.2014.6957752.

[41] M. Soilán, B. Riveiro, J. Martínez-Sánchez, and P. Arias, "Traffic sign detection in MLS acquired point clouds for geometric and image-based semantic inventory," *ISPRS J. Photogramm. Remote Sens.*, 2016, doi: 10.1016/j.isprsjprs.201 6.01.019.

[42] R. Schram, A. Williams, M. van Ratingen, J. Strandroth, A. Lie, and M. Paine, "New NCAP test and assessment protocols for speed assistance systems, a first in many ways," in *23rd Int. Tech. Conf. Enhanced Safety Veh. (ESV): Research Collaboration to Benefit Safety of All Road Users*, Seoul, South Korea, 2013.

[43] Euro NCAP, "Assessment Protocol - Overall Rating v9.0.1," *Euro NCAP Protoc.*, 2020.

[44] G. Piccioli, E. De Micheli, P. Parodi, and M. Campani, "Robust method for road sign detection and recognition," *Image Vis. Comput.*, vol. 14, no. 3, pp. 209–223, Apr. 1996, doi: 10.1016/0262-8856(95)01057-2.

[45] S. Maldonado-Bascon, S. Lafuente-Arroyo, P. Gil-Jimenez, H. Gomez-Moreno, and F. Lopez-Ferreras, "Road-sign detection and recognition based on support vector machines,"

IEEE Trans. Intell. Transp. Syst., vol. 8, no. 2, pp. 264–278, Jun. 2007, doi: 10.1109/TITS.2007.895311.

[46] F. Zaklouta and B. Stanciulescu, "Real-time traffic sign recognition in three stages," Rob. Auton. Syst., vol. 62, no. 1, pp. 16–24, Jan. 2014, doi: 10.1016/j.robot.2012.07.019.

[47] J. Stallkamp, M. Schlipsing, J. Salmen, and C. Igel, "The German traffic sign recognition benchmark: A multi-class classification competition," in Proceedings of the International Joint Conference on Neural Networks, 2011, doi: 10.1109/IJCNN.2011 .6033395.

[48] F. Larsson and M. Felsberg, "Using Fourier descriptors and spatial models for traffic sign recognition," Lecture Notes in Computer Science (including subseries Lecture Notes in Artificial Intelligence and Lecture Notes in Bioinformatics), 2011, doi: 10.1 007/978-3-642-21227-7_23.

[49] A. Mogelmose, D. Liu, and M. M. Trivedi, "Detection of U.S. traffic signs," IEEE Trans. Intell. Transp. Syst., vol. 16, no. 6, pp. 3116–3125, Dec. 2015, doi: 10.1109/TITS.2015.2433019.

[50] U. Nations, "No. 16743. Convention on road signs and signals. Concluded at Vienna on 8 November 1968," 1998, pp. 466–466.

[51] NHTSA, "Distracted driving in fatal crashes, 2017," 2019. doi: DOT HS 811 379.

[52] World Health Organization (WHO), "Mobile phone use: A growing problem of driver distraction," Technology, 2011.

[53] Euro NCAP, "Euro NCAP 2025 roadmap: In pursuit of Vision Zero," Euro NCAP Rep., 2017.

[54] T. Ranney, E. Mazzae, R. Garrott, and M. Goodman, "NHTSA driver distraction research: Past, present, and future," USDOT, Natl. Highw. Traffic Saf. Adm., 2000.

[55] J. L. Hossain, P. Ahmad, L. W. Reinish, L. Kayumov, N. K. Hossain, and C. M. Shapiro, "Subjective fatigue and subjective sleepiness: Two independent consequences of sleep disorders?," J. Sleep Res., 2005, doi: 10.1111/j.1365-2869.2 005.00466.x.

[56] B. Fischler, "Review of clinical and psychobiological dimensions of the chronic fatigue syndrome: Differentiation from

depression and contribution of sleep dysfunctions," *Sleep Med. Rev.*, 1999, doi: 10.1016/S1087-0792(99)90020-5.

[57] T. J. Balkin and N. J. Wesensten, "Differentiation of slee-piness and mental fatigue effects," in *Cognitive Fatigue: Multidisciplinary Perspectives on Current Research and Future Applications*, 2010.

[58] NHTSA, "Visual-manual NHTSA driver distraction guide-lines for in-vehicle electronic devices," Docket No. NHTSA-2010-0053, 2013.

[59] P. Viola and M. Jones, "Rapid object detection using a boosted cascade of simple features," *Proceedings of the IEEE Computer Society Conference on Computer Vision and Pattern Recognition*, 2001, doi: 10.1109/cvpr.2001.990517.

[60] L. T. Nguyen-Meidine, E. Granger, M. Kiran, and L. A. Blais-Morin, "A comparison of CNN-based face and head detec-tors for real-time video surveillance applications," *Proceedings of the 7th International Conference on Image Processing Theory, Tools and Applications*, IPTA 2017, 2018, doi: 10.1109/IPTA.2017.831 0113.

[61] S. El Kaddouhi, A. Saaidi, and M. Abarkan, "Eye detection based on the Viola-Jones method and corners points," *Multimed. Tools Appl.*, 2017, doi: 10.1007/s11042-017-4415-5.

[62] K. Kircher and C. Ahlstrom, "Issues related to the driver distraction detection algorithm AttenD," in *1st Int. Conf. Driv. Distraction Ina.*, 2009.

[63] A. E. A. M. Association and others, "ACEA Pocket Guide 2019--2020," p. 53, 2020.

[64] R. A. Bolt, "'Put-that-there,'" *ACM SIGGRAPH Comput. Graph.*, vol. 14, no. 3, pp. 262–270, Jul. 1980, doi: 10.1145/9651 05.807503.

[65] C. Van Nimwegen and K. Schuurman, "Effects of gesture-based interfaces on safety in automotive applications," *Adjun. Proc. - 11th Int. ACM Conf. Automot. User Interfaces Interact. Veh. Appl. AutomotiveUI 2019*, pp. 292–296, 2019, doi: 10.1145/3349263.3351522.

[66] T. Lewin, "The BMW century: The ultimate performance machines," *Motorbooks International*, p. 225, 2016.

[67] M. Zobl, R. Nieschulz, M. Geiger, M. Lang, and G. Rigoll, "Gesture components for natural interaction with in-car devices," *Lect. Notes Artif. Intell. (Subseries Lect. Notes Comput. Sci.)*, vol. 2915, pp. 448–459, 2004, doi: 10.1007/978-3-540-24598-8_41.

[68] K. A. Smith, C. Csech, D. Murdoch, and G. Shaker, "Gesture recognition using mm-wave sensor for human-car interface," *IEEE Sensors Lett.*, vol. 2, no. 2, pp. 1–4, Jun. 2018, doi: 10.1109/LSENS.2018.2810093.

[69] P. Molchanov, S. Gupta, K. Kim, and K. Pulli, "Multi-sensor system for driver's hand-gesture recognition," *2015 11th IEEE International Conference and Workshops on Automatic Face and Gesture Recognition (FG)*, May 2015, pp. 1–8, doi: 10.1109/FG.2015.7163132.

[70] S. S. Rautaray and A. Agrawal, "Vision based hand gesture recognition for human computer interaction: a survey," *Artif. Intell. Rev.*, vol. 43, no. 1, pp. 1–54, Jan. 2015, doi: 10.1007/s10462-012-9356-9.

[71] C. Pearl, *Designing Voice User Interfaces: Principles of Conversational Experiences*, 1st ed. O'Reilly Media, Inc., 2016.

[72] Y. Lin, J.-B. Michel, E. Lieberman Aiden, J. Orwant, W. Brockman, and S. Petrov, "Syntactic annotations for the google books ngram corpus," Jeju, Repub. Korea, 2012.

[73] G. D. Forney, "The Viterbi algorithm," *Proc. IEEE*, 1973, doi: 10.1109/PROC.1973.9030.

[74] Daimler, "World premiere at CES 2018: MBUX: A completely new user experience for the new compact cars," Press Release, 2018, [Online]. Available: https://media.daimler.com/marsMediaSite/ko/en/32705627.

[75] G. Tecuci, D. Marcu, M. Boicu, and D. A. Schum, "Introduction," in *Knowledge Engineering*, Cambridge: Cambridge University Press, 2016, pp. 1–45.

[76] R. Burke, "Hybrid recommender systems: Survey and experiments hybrid recommender systems," *Res. Gate*, 2016, doi: 10.1023/A.

[77] M. Unger, A. Tuzhilin, and A. Livne, "Context-aware recommendations based on deep learning frameworks," *ACM*

Trans. Manage. Inf. Syst., vol. 11, no. 2, Article 8, July 2020, https://doi.org/10.1145/3386243.

[78] S. Zhang, L. Yao, A. Sun, and Y. Tay, "Deep learning based recommender system: A survey and new perspectives," ACM Computing Surveys., 2019, doi: 10.1145/3285029.

[79] S. Halder, A. Ghosal, and M. Conti, "Secure over-the-air software updates in connected vehicles: A survey," Comput. Networks, vol. 178, p. 107343, 2020, doi: 10.1016/j.comnet.2020.107343.

[80] D. Shapiro, "Mercedes-Benz, NVIDIA partner to build the world's most advanced, software-defined vehicles," 2020. https://blogs.nvidia.com/blog/2020/06/23/mercedes-benz-nvidia-software-defined-vehicles/ (accessed Jun. 24, 2020).

[81] M. Ziegler, S. Rossmann, A. Steer, and S. Danzer, "Leading the way to an AI-driven organization - Porsche Consulting," pp. 1–35, 2019, [Online]. Available: https://www.porsche-consulting.com/fileadmin/docs/04_Medien/Publikationen/SRX04107_AI-driven_Organizations/Leading_the_Way_to_an_AI-Driven_Organization_2019_C_Porsche_Consulting-v2.pdf.

[82] T. Fountaine, B. McCarthy, and T. Saleh, "Building the AI-powered organization," HBR, no. August, 2019.

[83] F. Lambert, "Tesla has now 1.3 billion miles of Autopilot data going into its new self-driving program," Electrek, 2016. https://electrek.co/2016/11/13/tesla-autopilot-billion-miles-data-self-driving-program/ (accessed May 30, 2020).

[84] H. Hourani and A. Hammad, "The Impact of artificial intelligence on software testing," 2019 IEEE Jordan Int. Jt. Conf. Electr. Eng. Inf. Technol. JEEIT 2019 - Proc., pp. 565–570, 2019.

[85] C. E. Tuncali, G. Fainekos, H. Ito, and J. Kapinski, "Simulation-based adversarial test generation for autonomous vehicles with machine learning components," IEEE Intell. Veh. Symp. Proc., vol. 2018-June, no. Iv, pp. 1555–1562, 2018, doi: 10.1109/IVS.2018.8500421.

[86] F. Lambert, "Tesla's fleet has accumulated over 1.2 billion miles on Autopilot and even more in 'shadow mode', report

says - Electrek," *Electrek*, 2018. https://electrek.co/2018/07/1 7/tesla-autopilot-miles-shadow-mode-report/ (accessed May 21, 2020).

[87] Y. Kang, H. Yin, and C. Berger, "Test your self-driving algorithm: An overview of publicly available driving datasets and virtual testing environments," *IEEE Trans. Intell. Veh.*, vol. 4, no. 2, pp. 171–185, 2019, doi: 10.1109/TIV.201 8.2886678.

[88] F. Rosique, P. J. Navarro, C. Fernández, and A. Padilla, "A systematic review of perception system and simulators for autonomous vehicles research," *Sensors (Switzerland)*, vol. 19, no. 3, pp. 1–29, 2019, doi: 10.3390/s19030648.

[89] J. Hanhirova, A. Debner, M. Hyyppä, and V. Hirvisalo, "A machine learning environment for evaluating autonomous driving software," 2020, [Online]. Available: http:// arxiv.org/abs/2003.03576.

[90] W. Li, "AADS: Augmented autonomous driving simulation using data-driven algorithms," *Sci. Robot.*, vol. 4, no. 28, 2019. https://robotics.sciencemag.org/content/4/28/eaaw0863/tab-pdf

[91] Q. Chao et al., "A survey on visual traffic simulation: Models, evaluations, and applications in autonomous driving," *Comput. Graph. Forum.*, vol. 39, no. 1, pp. 287–308, 2019, doi: 10.1111/cgf.13803.

[92] I. J. Goodfellow et al., "Generative adversarial nets," in *Advances in Neural Information Processing Systems*, 2014.

[93] M. Y. Liu, T. Breuel, and J. Kautz, "Unsupervised image-to-image translation networks," in *Advances in Neural Information Processing Systems*, 2017.

[94] X. Ouyang, Y. Cheng, Y. Jiang, C.-L. Li, and P. Zhou, "Pedestrian-synthesis-GAN: Generating pedestrian data in real scene and beyond," *CoRR*, vol. abs/1804.0, 2018, [Online]. Available: http://arxiv.org/abs/1804.02047

[95] T. Truong and S. Yanushkevich, "Generative Adversarial Network for Radar Signal Synthesis," in *Proceedings of the International Joint Conference on Neural Networks*, 2019, doi: 10.11 09/IJCNN.2019.8851887.

[96] L. Caccia, H. Van Hoof, A. Courville, and J. Pineau, "Deep generative modeling of LiDAR data," in *IEEE International Conference on Intelligent Robots and Systems*, 2019, doi: 10.1109/IROS40897.2019.8968535.

[97] R. Meyes, J. Donauer, A. Schmeing, and T. Meisen, "A recurrent neural network architecture for failure prediction in deep drawing sensory time series data," *Procedia Manuf.*, vol. 34, pp. 789–797, 2019, doi: 10.1016/j.promfg.2019.06.205.

[98] H. Khayyam, B. Javadi, M. Jalili, and R. N. Jazar, "Artificial intelligence and internet of things for autonomous vehicles," *Nonlinear Approaches Eng. Appl.*, pp. 39–68, 2020, doi: 10.1007/978-3-030-18963-1_2.

[99] A. Nikitas, K. Michalakopoulou, E. T. Njoya, and D. Karampatzakis, "Artificial intelligence, transport and the smart city: Definitions and dimensions of a new mobility era," *Sustain.*, vol. 12, no. 7, pp. 1–19, 2020, doi: 10.3390/su12072789.

[100] R. Abduljabbar, H. Dia, S. Liyanage, and S. A. Bagloee, "Applications of artificial intelligence in transport: An overview," *Sustain.*, vol. 11, no. 1, pp. 1–24, 2019, doi: 10.3390/su11010189.

[101] R. Prytz, "Machine learning methods for vehicle predictive maintenance using off-board and on-board data," no. 9. 2014.

[102] M. Uricar, P. Krizek, G. Sistu, and S. Yogamani, "SoilingNet: Soiling detection on automotive surround-view cameras," *2019 IEEE Intell. Transp. Syst. Conf. ITSC 2019*, pp. 67–72, 2019, doi: 10.1109/ITSC.2019.8917178.

[103] M. Woschank, E. Rauch, and H. Zsifkovits, "A review of further directions for artificial intelligence, machine learning, and deep learning in smart logistics," *Sustain.*, vol. 12, no. 9, 1–23, 2020, doi: 10.3390/su12093760.

[104] Y. Ran, X. Zhou, P. Lin, Y. Wen, and R. Deng, "A survey of predictive maintenance: Systems, purposes and approaches," vol. XX, no. Xx, pp. 1–36, 2019, [Online]. Available: http://arxiv.org/abs/1912.07383.

[105] U. Shafi, A. Safi, A. R. Shahid, S. Ziauddin, and M. Q. Saleem, "Vehicle remote health monitoring and prognostic maintenance system," J. *Adv. Transp.*, vol. 2018, pp. 1–10, 2018, doi: 10.1155/2018/8061514.

[106] V. F. A. Meyer, "Challenges and reliability of predictive maintenance," no. March, p. 16, 2019, doi: 10.13140/RG.2.2.35379.89129.

[107] K. Korishchenko, I. Stankevich, N. Pilnik, and D. Petrova, "Usage-based vehicle insurance: Driving style factors of accident probability and severity," 2019, [Online]. Available: http://arxiv.org/abs/1910.00460.

[108] S. Arumugam and R. Bhargavi, "A survey on driving behavior analysis in usage based insurance using big data," J. *Big Data*, vol. 6, no. 1, pp. 1–21, 2019, doi: 10.1186/s40537-019-0249-5.

[109] M. Soleymanian, C. B. Weinberg, and T. Zhu, "Sensor data and behavioral tracking: Does usage-based auto insurance benefit drivers?," *Mark. Sci.*, vol. 38, no. 1, pp. 21–43, 2019, doi: 10.1287/mksc.2018.1126.

[110] J. F. Júnior et al., "Driver behavior profiling: An investigation with different smartphone sensors and machine learning," *PLoS One*, vol. 12, no. 4, pp. 1–16, 2017, doi: 10.1371/journal.pone.0174959.

[111] Z. Chen, J. Yu, Y. Zhu, Y. Chen, and M. Li, "D3: Abnormal driving behaviors detection and identification using smartphone sensors," in *2015 12th Annu. IEEE Int. Conf. Sensing, Commun. Networking*, SECON 2015, pp. 524–532, 2015, doi: 10.1109/SAHCN.2015.7338354.

[112] H. Chen and L. Chen, "Support vector machine classification of drunk driving behaviour," *Int. J. Environ. Res. Public Health*, vol. 14, no. 1, pp. 1–14, 2017, doi: 10.3390/ijerph14010108.

[113] J. Zhang et al., "A deep learning framework for driving behavior identification on in-vehicle CAN-BUS sensor data," *Sensors (Switzerland)*, vol. 19, no. 6, pp. 6–8, 2019, doi: 10.3390/s19061356.

[114] S. Shalev-Shwartz and A. Shashua, "On the sample complexity of end-to-end training vs. semantic abstraction

training," pp. 1–4, 2016, [Online]. Available: http://arxiv.org/abs/1604.06915.

[115] M. Bojarski et al., "End to end learning for self-driving cars," pp. 1–9, 2016, [Online]. Available: http://arxiv.org/abs/1604.07316.

[116] M. Bojarski et al., "Explaining how a deep neural network trained with end-to-end learning steers a car," pp. 1–8, 2017, [Online]. Available: http://arxiv.org/abs/1704.07911.

[117] F. Codevilla, M. Miiller, A. Lopez, V. Koltun, and A. Dosovitskiy, "End-to-end driving via conditional imitation learning," in *Proc. IEEE Int. Conf. Robot. Autom.*, pp. 4693–4700, 2018, doi: 10.1109/ICRA.2018.8460487.

[118] A. El Sallab et al., "End-to-end multi-modal sensors fusion system For urban automated driving," NIPS 2018 Work. MLITS Submiss., no. CoRL, 2018.

[119] J. Kocić, N. Jovičić, and V. Drndarević, "An end-to-end deep neural network for autonomous driving designed for embedded automotive platforms," *Sensors*, vol. 19, no. 9, pp. 1–26, 2019.

[120] S. Hecker, D. Dai, and L. Van Gool, "End-to-end learning of driving models with surround-view cameras and route planners," *Lect. Notes Comput. Sci. (including Subser. Lect. Notes Artif. Intell. Lect. Notes Bioinformatics)*, vol. 11211 LNCS, pp. 449–468, 2018, doi: 10.1007/978-3-030-01234-2_27.

[121] T. Glasmachers, "Limits of end-to-end learning," *J. Mach. Learn. Res.*, vol. 77, pp. 17–32, 2017.

[122] S. Shalev-Shwartz and A. Shashua, "On the sample complexity of end-to-end training vs. semantic abstraction training," pp. 1–4, 2016.

[123] Euro NCAP, "The ratings explained," 2020. https://www.euroncap.com/en/vehicle-safety/the-ratings-explained/ (accessed Aug. 08, 2020).

[124] I. O. for Standardization, "Road vehicles — Safety of the intended functionality (Standard No. 21448:2019)," 2019. https://www.iso.org/standard/70939.html (accessed Aug. 08, 2020).

[125] T. Stolte, G. Bagschik, and M. Maurer, "Safety goals and functional safety requirements for actuation systems of automated vehicles," *IEEE Conf. Intell. Transp. Syst. Proceedings, ITSC*, pp. 2191–2198, 2016, doi: 10.1109/ITSC.201 6.7795910.

[126] N. O'Mahony et al., "Deep learning vs. traditional computer vision," *Adv. Intell. Syst. Comput.*, vol. 943, no. Cv, pp. 128–144, 2020, doi: 10.1007/978-3-030-17795-9_10.

[127] M. Gharib, P. Lollini, M. Botta, E. Amparore, S. Donatelli, and A. Bondavalli, "On the safety of automotive systems incorporating machine learning based components: A position paper," *Proc. - 48th Annu. IEEE/IFIP Int. Conf. Dependable Syst. Networks Work. DSN-W 2018*, pp. 271–274, 2018, doi: 10.1109/DSN-W.2018.00074.

[128] Z. El-Rewini, K. Sadatsharan, D. F. Selvaraj, S. J. Plathottam, and P. Ranganathan, "Cybersecurity challenges in vehicular communications," *Veh. Commun.*, vol. 23, p. 100214, 2020, doi: 10.1016/j.vehcom.2019.100214.

[129] P. Sewalkar and J. Seitz, "Vehicle-to-pedestrian communication for vulnerable road users: Survey, design considerations, and challenges," *Sensors (Switzerland)*, vol. 19, no. 2:358, pp. 1–18, 2019, doi: 10.3390/s19020358.

[130] B. Klein, "Cybersecurity in automotive," no. March, 2020.

[131] D. Kleinz, "Automotive Cybersecurity - ISO/SAE 21434," 2020. https://www.linkedin.com/pulse/automotive-cybersecurity-isosae-21434-david-kleinz-cissp/?trackingId=5hoE66rld6CgA nmumnaxRQ%3D%3D.

[132] F. Sommer, J. Dürrwang, and R. Kriesten, "Survey and classification of automotive security attacks," *Inf.*, vol. 10, no. 4:148, pp. 1–29, 2019, doi: 10.3390/info10040148.

[133] E. Aliwa, O. Rana, and P. Burnap, "Cyberattacks and countermeasures for in-vehicle networks," *arXiv Prepr. arXiv2004.10781*, pp. 1–37, 2020.

[134] D. S. Berman, A. L. Buczak, J. S. Chavis, and C. L. Corbett, "A survey of deep learning methods for cyber security," *Inf.*, vol. 10, no. 4:122, pp. 1–35, 2019, doi: 10.3390/info10040122.

[135] N. Kaloudi and L. I. Jingyue, "The AI-based cyber threat landscape: A survey," *ACM Comput. Surv.*, vol. 53, no. 1, pp. 1–34, 2020, doi: 10.1145/3372823.

[136] Z. Khan, M. Chowdhury, M. Islam, C.-Y. Huang, and M. Rahman, "Long short-term memory neural networks for false information attack detection in software-defined in-vehicle network," 2019, [Online]. Available: http://arxiv.org/abs/1906.10203.

[137] N. O. Leslie, C. A. Kamhoua, and C. S. Tucker, "Generative adversarial attacks against intrusion detection systems using active learning," pp. 1–6.

[138] A. Tuor, S. Kaplan, B. Hutchinson, N. Nichols, and S. Robinson, "Deep learning for unsupervised insider threat detection in structured cybersecurity data streams," *AAAI Work. - Tech. Rep.*, vol. WS-17-01-, no. 2012, pp. 224–234, 2017.

[139] M. Bouton, "Utility decomposition for planning under uncertainty for autonomous driving," *Proc. Int. Jt. Conf. Auton. Agents Multiagent Syst. AAMAS*, vol. 3, pp. 1731–1732, 2018.

INDEX

Printed in the United States
by Baker & Taylor Publisher Services

Printed in the United States
by Baker & Taylor Publisher Services